HYDRAULICS AND HYDROLOGY FOR STORMWATER MANAGEMENT

John E. Gribbin, P.E.
Essex County College

Delmar Publishers

I⟨T⟩P™ an International Thomson Publishing Company

Albany • Bonn • Boston • Cincinnati • Detroit • London • Madrid •
Melbourne • Mexico City • New York • Pacific Grove • Paris •
San Francisco • Singapore • Tokyo • Toronto • Washington

NOTICE TO THE READER

Delmar Staff
Acquisitions Editor: John Anderson
Developmental Editor: Michelle Ruelos Cannistraci
Production Coordinator: Karen Smith
Art/Design Coordinator: Cheri Plasse

COPYRIGHT © 1997
By Delmar Publishers
a division of International Thomson Publishing Inc.

The ITP logo is a trademark under license.

Printed in the United States of America

For more information, contact:

Delmar Publishers
3 Columbia Circle, Box 15015
Albany, New York 12212-5015

International Thomson Publishing Europe
Berkshire House 168-173
High Holborn
London WC1V 7AA
England

Thomas Nelson Australia
102 Dodds Street
South Melbourne, 3205
Victoria, Australia

Nelson Canada
1120 Birchmont Road
Scarborough, Ontario
Canada M1K 5G4

International Thomson Editores
Campos Eliseos 385, Piso 7
Col Polanco
11560 Mexico D F Mexico

International Thomson Publishing GmbH
Königswinterer Strasse 418
53227 Bonn
Germany

International Thomson Publishing Asia
221 Henderson Road
#05-10 Henderson Building
Singapore 0315

International Thomson Publishing—Japan
Hirakawacho Kyowa Building, 3F
2-2-1 Hirakawacho
Chiyoda-ku, Tokyo 102
Japan

2 3 4 5 6 7 8 9 10 XXX 01 00 99 98

Library of Congress Cataloging-in-Publication Data

Gribbin, John E.
    Hydraulics and hydrology for stormwater management / by John E. Gribbin.
        p.   cm.
    Includes bibliographical references and index.
    ISBN 0-8273-7278-7
    1. Urban runoff—Management.   2. Hydraulics.   3. Hydrology.
TD657.G75   1996
628'.21—dc20   96-18771
                                                                CIP

Delmar Publishers' Online Services

To access Delmar on the World Wide Web, point your browser to:

http://www.delmar.com/delmar.html

To access through Gopher: gopher://gopher.delmar.com

(Delmar Online is part of "thomson.com", an Internet site with information on more than 30 publishers of the International Thomson Publishing organization.)

For information on our products and services:
email: info @ delmar.com
or call 800-347-7707

In memory of my father,
John B. Gribbin,
Associate Professor of Civil Engineering at Manhattan College.

ACKNOWLEDGMENTS

The author wishes to express appreciation to the following individuals for their helpful review and contributions to the text:

James Carr
Northwest Mississippi Community College
Senatobia, MS 38668

Harold Dendy
Itawamba Community College
Tupelo, MS

Kenneth Driver
A. B. Technical Community College
Asheville, NC 28801

David J. Hanna
Ferris State University
Big Rapids, MI

Professor H. Carlton Moor
Wentworth Institute of Technology
Boston, MA 02115

Alfred Pagan, P.E.
Consulting Engineer
Westwood, NJ

Charles Partlow
SUNY - Canton
Canton, NY

Charles Scribner
Vermont Technical College
Randolph Center, VT 05061

Professor Erdogan Sener
Purdue University
Indianapolis, IN

Donald Woodward, P.E.
NRCS
Washington, DC

Randal Tweden
Dunwoody Institute
Minneapolis, MN 55403-1192

Also, I thank my son Peter for all his tireless typing and the rest of my family for their encouragement.

CONTENTS

PREFACE

This text has been written to fill a long-standing need to treat the topics of stormwater runoff and hydraulics together in one book. It is intended to be used by students of civil engineering technology and construction technology as well as practitioners in industry and government. The topics presented are relevant to public works, land development and municipal engineering and planning; in fact, to any designer (both engineer and technician) who must deal with the conveyance of stormwater in any aspect of his/her work.

The book includes several features designed to make the learning process more accessible and streamlined, such as:

- Many easy to follow examples

- Many clear diagrams, charts and topographic maps to illustrate concepts developed in the text

- Case studies based upon real-world projects

- A list of objectives starting each chapter to help focus the attention of the reader

- Actual design charts in the Appendix to enable examples and problems to relate to real situations

- A list of current computer software in the Appendix to enable the reader to investigate this essential tool

- A comprehensive glossary to help the reader understand important terms

The subjects of hydraulics and hydrology include many more topics than those presented in this text. Hydraulics texts are available which treat engineering hydraulics in a comprehensive manner and likewise there are hydrology texts dealing only with the engineering aspects of hydrology. But this text pares down the many aspects of hydraulics and hydrology to the most basic and common areas dealing with stormwater management encountered by the designer on a day to day basis.

Principle topics include:

- Background concepts such as historical overview and basic notions of computation and design

- Fundamental hydrostatics

- Fundamental hydrodynamics

- Open channel hydraulics

- Runoff computations (Rational & SCS Methods)

- Design of culverts, storm sewers, detention basins and outflow protection

One of the outstanding features of the book is the treatment of runoff computations. Thorough analysis and practice of watershed delineation is included to hone this skill which is essential to runoff analysis but often lacking in designers' training.

Another outstanding feature of the text is the comprehensive appendix which includes excerpts from several relevant design manuals in use today. Students and others using the text will continually refer to the design charts located in Appendix A through Appendix D when studying examples and working problems. Mastering the use of the charts is an indispensible benefit in learning the techniques of problem solving in the real world. The student will learn not only the use of the charts but also the theory and rationale used to create them.

For example, when analyzing a culvert problem, the student learns to recognize the correct chart in Appendix B and then utilizes it to derive key numerical values needed for the problem's solution. References to specific Appendix sections are included throughout the text to help guide the reader in their proper use.

In addition to the development of hydraulic theory and runoff computation techniques, one of the goals of the text is to introduce the reader to some of the rudimentary stormwater management design processes used in civil engineering practice. To accomplish this, realistic design problems and case studies are included which rely upon actual design charts. However, the text should not be construed as a complete design manual to be used on the job, nor is it intended to be. Good engineering practice normally requires the use of a variety of comprehensive sources found in professional publications and design manuals prepared by governmental agencies.

In developing the various topics throughout the text, the author has assumed certain prior knowledge on the part of the reader. This includes fundamental concepts of land surveying, interpretation of topographic maps, profiles and cross sections. Also, such other concepts as principles of engineering graphics and the use of the engineer's scale are prerequisites to a full understanding of the text.

C H A P T E R

1

HYDRAULICS AND HYDROLOGY IN ENGINEERING

Engineers cannot avoid confronting the problems posed by rainfall and its consequent runoff. Some of mankind's earliest endeavors centered on this age-old battle with the forces of nature in the form of water. For most of our history, the engineer and his predecessors dealt with water problems by utilizing various rule-of-thumb solutions, that is, whatever seemed to work. Only in recent times have these endeavors taken on a systematic body of laws and quantitative formulae.

In this chapter, we will take a brief trip through the world of engineering hydraulics and hydrology of yesterday and today. We will look at the various aspects of modern stormwater management and also review some general principles of engineering design.

OBJECTIVES

After completing this chapter, the reader should be able to:
- Place hydraulic/hydrologic engineering in a historical perspective
- Define stormwater management
- Explain the roles of public agencies in stormwater management
- Appreciate the factors involved in engineering design
- Perform computations using the appropriate significant figures
- Convert computations to metric units when necessary

1.1 History of Water Engineering

People first started manipulating water on a large scale as a response to the need for irrigation tied to the first attempts to form an agrarian society. The first known large scale irrigation project was undertaken in Egypt approximately five thousand years ago. In the following millennia, many other water projects sprang up in the Mediterranean and Near Eastern worlds. These included dams, canals, aqueducts and sewer systems. The conveyance of water through pipes was also developed in ancient times. In China, bamboo pipes were used as early as 2500 B.C. and the Romans utilized lead and bronze pipes by 200 B.C.

Figure 1-1 The Pont du Gard, a three-level Roman aqueduct bridge near Nîmes, France.

The Romans' prowess as engineers was amply demonstrated in their hydraulic systems. The famous aqueducts were among the wonders of the world and remained in use through two millennia (see Figure 1-1). The Greeks, although not the engineers that the Romans were, nonetheless made significant contributions to the theories of hydraulics. Archimedes is considered the earliest contributor to hydraulics based upon truly scientific work. In about 250 B.C., he published a written work on hydrostatics which presented the laws of buoyancy (Archimedes Principle) and flotation. He is generally considered the father of hydrostatics.

During the period from 500 B.C. into the Middle Ages, irrigation and water supply systems were constructed and maintained in such diverse locations as China, the Roman Empire and North America. Such engineering was designed and constructed by artisans using rules of thumb, artisans who, despite Archimedes work, lacked the benefits of scientific inquiry. The great Roman engineers, for example, did not understand the concept of velocity, and it was not until 1500 A.D. that the connection between rainfall and streamflow was taken seriously.

As the Roman Empire declined, many of the advances made during the Greco-Roman period were forgotten, only to be "rediscovered" during the Renaissance. It was during this period that hydraulics began to be developed as a science.

The first effort at organized engineering knowledge was the founding in 1760 of the École des Ponts et Chaussées in Paris. In 1738, Daniel Bernoulli published his famous *Bernoulli Equation* formulating the conservation of energy in hydraulics. During the eighteenth and nineteenth centuries, advances in hydraulic engineering laid the groundwork for further developments in the twentieth century and is referred to as the classical period of hydraulics.

Despite the domination of the French during the classical period, work was conducted in other countries as well. In England, for instance, John Smeaton was very active in many aspects of hydraulic engineering and was the first to call himself a *civil engineer*.

As late as 1850, however, engineering designs were still based mainly on rules of thumb developed through experience and tempered with liberal factors of safety. Since that time, utilization of theory has increased rapidly. Today a vast amount of careful computation is an integral part of most project designs. Figure 1-2 depicts one of many hydraulic experiments conducted in Lowell, Massachusetts, in the mid 1800s that contributed greatly to the field of hydraulic engineering.

Figure 1-2 The city of Lowell, Massachusetts, was the site of a famous series of water power experiments conducted by James Francis and Uriah Boyden. The publication of the results in 1855 contributed greatly to the field of hydraulic engineering. Here, two assistants are shown measuring water levels. *(Courtesy of University of Massachusetts Lowell, Locks and Canal Collection.)*

1.2 Modern Practice of Stormwater Management

Civil engineers work with water wherever it affects the structures and infrastructure of our civilization. The role of the civil engineer and technician in connection with the many diverse effects of water may be grouped into three broad categories:

1. Flood Control—managing the natural flow of stormwater to prevent property damage and loss of life.
2. Water Resources—exploiting the available water resources for beneficial purposes such as water supply, irrigation, hydro-electric power and navigation.
3. Water Quality—managing the use of water to prevent its degradation due to pollutants both natural and man-made.

Although the first role listed above, flood control, constitutes the primary focus of this text, the other two are of no less importance. All three areas constitute projects designed and carried out by people working in both the private and public sectors.

As an example of private endeavors in flood control, imagine that an entrepreneur wishes to construct a factory surrounded by a parking area. He/she must engage a civil engineer to design proper grading and a storm sewer system to adequately convey any rainfall occurring on the site. In addition, a detention basin may be required to prevent any adverse effect of runoff from the factory site to adjacent properties.

Even though these problems will be solved by an engineering firm contracting directly with the owner, public agencies become involved as well since all designs affecting the public welfare must be reviewed and approved by the appropriate local, county and state agency.

Examples of public endeavors in flood control are many and may be as simple as the design of a pipe culvert under a newly constructed road to allow free passage of a stream, or as involved as a large regional detention basin intended to reduce stormwater runoff quantity affecting a relatively large population. Each of these public projects may be designed by engineers employed by public agencies, or they may be designed by private engineers contracting directly with the appropriate public agency.

In a typical privately owned land development project, the engineer representing the developer works with the engineer representing the regulating agency to solve any stormwater runoff problems. The relationship between the engineers is at once adversarial and cooperative as they work to protect the respective interests of the private developer and of the public. In this way, they create the best possible project.

The term *stormwater management* as used in this text refers to the engineering practices and regulatory policies employed to mitigate the adverse effects of stormwater runoff. These endeavors usually are associated with runoff problems resulting from various types of land development.

1.3   Legal and Environmental Issues

Over the past three decades, legal and environmental issues have dramatically changed the way civil engineers practice their art, and hydraulic/hydrologic engineering is no exception. Stormwater management was once based upon the principles of *good engineering practice,* but now design must also satisfy a myriad of regulations enforced by several levels of public agencies.

When hydraulic and hydrologic design affects the public, there is a legal issue, and when it affects the environment, there is an environmental issue. These two issues usually overlap since anything that affects the environment most often affects the public. Although legal and environmental issues abound throughout all areas of civil engineering, we will only look at a few that affect stormwater management on a regular basis.

When rain falls from the sky, it strikes the earth and then runs downhill, impelled by gravity across the land as it seeks the streams and rivers that eventually carry it to the sea. All such motion of water is considered by our society as naturally occurring, and if the water does damage along its path, such as erosion or flooding, no legal blame is assigned to any person. But the minute people alter their land in such a way as to change the course of the stormwater, they become liable for any damage done as a result of the alteration. The two ways that land development generally affects downstream property is by concentrating the flow of stormwater and by increasing the quantity of that flow.

The practice of stormwater management must take these problems into consideration and mitigate them. Mitigation is achieved by a variety of methods including re-routing the flow, dispersing the flow, lining the ground with erosion protection and providing a detention basin, among others.

Another problem that occurs in hydraulic and hydrologic design is the pollution of stormwater. Development of the land can and usually does result in several unwanted pollutants mixing into the stormwater as it runs off the developed site. These include salts and oils from paved areas, or fertilizer, pesticides and silt particles from vegetated areas. Stormwater management mitigates these problems by such measures as providing vegetative filters, siltation basins, catch basins and recharge basins.

Wetlands are an environmental feature which has come into prominence throughout the past decade. Wetlands are naturally occurring areas of land retaining water throughout much of the year. They are beneficial to the ecosystem and are particularly sensitive to disruption by the effects of development. Extra care must be taken to identify, delineate and protect these areas when they are on or adjacent to a land development project.

Design engineers work hand in hand with regulators in addressing and solving the problems raised by legal and environmental issues. In the following section, we will briefly outline the identities of the regulating bodies and the roles they perform.

1.4   Public Agencies

Over the past few decades, life in the United States has become much more regulated than in previous times. Naturally, the field of civil engineering design has acquired its own array of specialized rules. The trend toward increased regulation has generally relied upon the concept that the government may regulate anything that affects the health and welfare of the public. In civil engineering, this means just about everything may be regulated.

Regulating began at the local level with the proliferation of zoning ordinances by municipalities everywhere. Zoning has become more and more complex over the years, especially over the past two decades. Originally the concept was simple: a factory, for instance, cannot be constructed in an area of town reserved for residential development. Today zoning regulates not only the type of development but also details of the infrastructure, such as pavement thicknesses for roads, key parameters in storm sewer design, and detailed methodology for detention design.

Local municipal zoning and land development ordinances are administered by the municipal engineer, a civil engineer employed by or under contract to the municipality; and the zoning officer (also called the building inspector), an employee of the municipality trained in building code enforcement.

One level above municipal is the county in which the project is located. County governments have jurisdiction over certain roads designated as *county roads* and usually most of the bridges and culverts in the county, except those under state highways. Thus, if the stormwater flowing from a project flows onto a county road or through a county culvert, the county engineering department has limited regulatory power over the project. In addition, the Natural Resources Conservation Service (NRCS), formerly called the Soil Conservation Service, regulates soil erosion aspects of development work through local offices around the country. In many areas, these are called Soil Conservation District (SCD) offices and are semi-autonomous regulatory agencies using regulations promulgated by the NRCS, a federal agency.

The next level of regulation consists of a few specially created regional, semi-autonomous agencies having jurisdiction over certain geographical problem areas. They are usually associated with environmentally sensitive areas which require more intense scrutiny than the rest of the land to be developed.

The next level of regulation is the state. State laws governing land development are administered by specially created agencies such as the state Department of Environmental Protection, which typically has jurisdiction over any project that affects a state-regulated entity including, for instance, wetlands, streams and water fronts, or state-owned entities such as highways and culverts.

Finally, the most overreaching level of regulation is the federal government. Federal laws such as the 1972 Clean Water Act are administered by such federal agencies as the Environmental Protection Agency (EPA) and the U.S. Army Corps of Engineers. Other federal agencies which affect hydraulic and hydrologic design are the Bureau of Public Roads, a division of the Commerce Department, and the Federal Highway Administration, a division of the Department of Transportation. These agencies affect local everyday projects principally through the publication of design manuals developed through research.

1.5   Engineering Design

All engineers and technicians, whether they are electrical, mechanical or civil, are engaged in design. The civil designer works on projects which can be as daunting in scope as a 500-foot high dam complete with hydro-electric power station or as mundane as a concrete pipe laid in a trench.

Regardless of the size of the project, the design process involves the complete specification of every aspect of the structure so that it can then be constructed based upon the resulting specifications. That is, the engineer or technician must think of every detail of the structure and successfully convey his/her thoughts to the builder.

Design Process

In designing a structure, several important steps are required to transform an initial idea into a clear and fully developed document ready for construction. The example of a storm sewer pipe can be used to illustrate the general steps in performing a typical design:

1. **Concept**—First, determine the basic concept of the design. In this case, it is to convey stormwater from one location to another.

2. **Base Map**—Next, prepare a base map showing the topographic features of the project site together with any pertinent property boundaries. (A good base map is essential to the successful design process.)

3. **Design Development**—Next, sketch alternate layouts of the pipe on the base map. Also, research other factors affecting the design, such as soil conditions, structural loading on the pipe, potential interference with other subsurface utilities, etc.

4. **Calculations**—Next, perform appropriate engineering computations of key design quantities; in this case, the anticipated amount of stormwater to be conveyed by the pipe and the resulting pipe size. The calculations should be in written form and contain any assumptions made and should be checked by another designer.

5. **Prepare Drawings and Specifications**—Finally, prepare drawings showing the layout in plan and profile including any details and notes needed to adequately describe the structure for use by the builder in constructing the project. Include written specifications if necessary.

Design Outcome

Design is an endeavor that is enriched with experience. As more and more projects are completed, the good practitioner starts to acquire a deeper appreciation for the larger picture surrounding the design and weaves that broader perspective into his/her work. It is not enough to imagine only the proper functioning of the structure but other factors must also be taken into consideration, such as proper maintenance, cost, safety during construction, availability of materials and so on. Because the design process is such a complex and ever-growing intellectual endeavor, an exhaustive definition is virtually impossible. However, certain basic elements can be identified:

Design is the process of determining the complete specification of a structure so that it will:

1. Perform its intended function under all foreseeable circumstances without failing
2. Be able to be constructed at a cost within the budget of the owner
3. Be able to be easily and effectively maintained
4. Conform to all applicable local, county, state and federal laws and regulations
5. Not interfere with other structures or utilities that could be constructed in its vicinity in the future
6. Be able to be constructed in a safe manner
7. Remain intact and functional throughout its intended lifetime
8. Not present a safety hazard to the public throughout its lifetime
9. Not unduly degrade the environment either during construction or throughout its lifetime
10. Be aesthetically pleasing

Each structure must be designed using all of the factors listed above regardless of its apparent simplicity. In the later chapters of this text, we will see how to employ the principles of design listed here in some commonly encountered projects related to hydraulic and hydrologic engineering.

1.6   Engineering Computations

Almost all engineering designs require some computing of numbers. Although the use of calculators and computers make computing relatively easy, an understanding of certain basic principles of computing are important to a successful design process.

Significant Figures

The concept of *significant figures* should be familiar to anyone engaged in the various aspects of the design process. The number of significant figures of a quantity is the number of digits used to form the quantity (except for zeros under certain circumstances as explained below). Thus, the quantities 429, 1.02 and 0.00315 each have three significant figures. The zeros in the third example are not significant because they are only place holders. Zeros can also be used as place holders at the right end of a quantity where the quantity has no decimal point. Thus, the quantities 450, 1500 and 92,000 each have two significant figures. If, however, a decimal point is added to the end of such a quantity, the zeros become significant. Thus, the quantities 450., 1500. and 92,000. have three, four and five significant figures, respectively.

Numbers subjected to the rules of significant figures generally are quantities which have been measured. For instance, if the length of a pipe is measured as being 229 feet, it is said that the length was measured to the nearest foot and the measurement has three significant figures. However, a pipe measured as 229.0 feet was measured to the nearest tenth of a foot and the measurement has four significant figures. A pipe measured as 230 feet was measured to the nearest ten feet and the measurement has only two significant figures. However, a pipe measured as 230. feet was measured to the nearest one foot and the measurement has three significant figures.

Numbers not subject to the rules for significant figures are pure numbers which cannot vary to any extent. These numbers include counting numbers and assumed quantities. For example, in the formula $c = 2\pi r$, the number 2 is a counting number and therefore perfectly precise. It is the same as if it was expressed as 2.00000. Also, if the radius, r, is assumed hypothetically to be 4 feet and not measured, then the quantity 4 is also perfectly precise and the same as if it was written 4.00000 feet. (Of course, if r is measured, it should be expressed with the number of significant figures corresponding to the precision of the measurement.)

The rules for computations are as follows:

1. *Multiplication and Division* The answer to a multiplication or division computation should have no more significant figures than the least number of significant figures in any quantity in the computation.

2. *Addition and Subtraction* The answer to an addition or subtraction computation should have no more digits to the right of the decimal point than the least number of digits to the right of the decimal point in any quantity in the computation.

3. *Computations in Series* If a series of computations is to be made where the answer to one is used as a quantity in the next, then only the final answer of the last computation should be rounded to significant figures. In such a case, the number of significant figures would be based upon all quantities used in all of the computations.

Example 1-1

Problem
Find the circumference of a pipe having a diameter measured to be 4.00 feet.

Solution
Since the diameter was measured to a precision of three significant figures, it is expressed with three significant figures. The formula for circumference is $c = \pi d$.

$c = \pi d$
$c = (\pi)(4.00)$
$c = 12.6$ ft   (Answer)

Although the calculator display shows 12.566371, only three significant figures can be used in the answer.

Note: If the measurement had been 4.0 feet, the computed circumference would be 13 feet.

Example 1-2

Problem
Find the circumference of a typical 4-foot diameter pipe.

Solution
In this case, the diameter is a theoretical value and not subject to significant figures.

$c = \pi d$
$c = (\pi)(4)$
$c = 12.566371$ ft   (Answer)

Although all of the figures shown above may be used in the answer, for practical reasons only three or four figures are generally used. If, however, the circumference is to be used in a further computation, as many figures as possible should be used and only the final answer would be subject to rounding to significant figures. See Example 1-3.

Example 1-3

Problem
Find the volume of a cylinder having a diameter measured as 2.3 feet and length measured as 8.25 feet.

Solution
First computation:

$a = \pi d^2/4$
$a = \pi(2.3)^2/4$
$a = 4.1547563$ ft^2 (shown on calculator)

Second computation:

$V = aL$
$V = (4.1547563)(8.25)$
$V = 34.276739$ ft^3 (shown on calculator)
$V = 34$ ft^3 (Answer)

The final answer is rounded to two significant figures because the quantity 2.3 in the first computation has two significant figures.

Note: If the first answer was rounded to two significant figures before performing the second computation, the final answer would have been 35 ft³.

Accuracy and Precision

One way to appreciate the importance of significant figures is to understand the distinction between accuracy and precision. The term *accuracy* means a value which is close to the actual value. Thus, if the actual length of a pipe is 230.0000 feet, then a length value of 231 feet is accurate within 0.4 percent. On the other hand, a length value of 232.15 feet is less accurate even though it is more precise. The term *precision* refers to the fineness of a measurement. A length of 232.15 feet is precise to within 0.01 feet; that is, it implies a length value somewhere between 232.145 feet and 232.155 feet. Although it is commonly assumed that the true length falls within this range of values, such an assumption is not necessarily correct. Precision does not ensure accuracy.

Sometimes it is tempting to think that increasing the precision of an answer will also increase its accuracy. Thus it is tempting to express the answer to Example 1-3 as 34.28 ft³ instead of 34 ft³. The true answer to Example 1-3 lies in a range between 33.5 ft³ and 34.5 ft³. So to state the answer as 34.28 ft³ is meaningless and misleading.

Engineering calculations must always be accurate. But the accuracy stems from careful measurements and correctly applied scientific principles, not from an overstatement of precision by writing too many significant figures in the answer.

Example 1-4

Problem
Find the cross-sectional area, a, of a circular pipe of diameter not known to you. (True diameter = 2.500 ft.)

Solution
1. Accurate solution: Suppose the diameter is measured as accurately as possible to be 2.4 ft.

$$a = \pi d^2/4$$
$$a = \pi(2.4)^2/4$$
$$a = 4.5 \text{ ft}^2 \quad \text{(Answer)}$$

The answer is reported with two significant figures because the value for d had two significant figures.

2. Precise solution: Suppose the diameter is measured sloppily to be 2.7 feet.

$$a = \pi d^2/4$$
$$a = \pi(2.7)^2/4$$
$$a = 5.726 \text{ ft}^2 \quad \text{(Answer)}$$

The answer is reported with four significant figures because the calculator displayed at least that many.

Although the *precise* solution looks more accurate, in fact it differs from the theoretical solution of 4.909 ft² by a greater percentage than does the *accurate* solution and therefore it is not acceptable despite its more precise appearance.

To ensure that computations have the highest degree of accuracy possible, certain techniques should always be employed.

1. Always try to check every computation or measurement by repeating the computation or measurement using an alternate method.

   For instance, if a distance between two stations on a baseline is computed mathematically to be 350 feet, check the answer graphically by measuring the distance on your map with a scale.

   Another example is the measurement of an area on a map. If the area is measured with a planimeter to be 2.35 acres, check the answer by roughly measuring a length and width with a scale and computing the rough area using a = L x W and comparing with measured area for approximate agreement.

2. Always check the computation by considering whether or not it seems reasonable.

   For instance, if the volume of a swimming pool is computed to be 150,000 ft^3, a little reflection on the number will reveal that it is far out of the realm of reason. At that point, the calculation should be reviewed for errors.

3. When practical, have another engineer or technician review the entire calculation for agreement with assumptions and computations.

Many computations are performed using computers. In most cases, these involve relatively inexpensive software loaded on a personal computer. However, before a computer can be used successfully, the principles of design and computation must be understood thoroughly.

Computers

The computer is a tool just as is the calculator and the engineer's scale and all the other design tools utilized by engineers. But it is important to remember that the computer is not a surrogate for sound engineering design judgment. It is essential to good engineering practice to resist the temptation to rely on the answers given by computers as if the software has the ability to make design judgments. All computer computations must be scrutinized and checked as would any other computation performed on a calculator or done longhand.

Design principles presented in the later chapters of this text employ the same methodology utilized by most computer software. Computer applications should be learned elsewhere *after* the governing principles are first acquired. For the reader's convenience, a selected list of computer software relevant to stormwater management problems is included in Appendix E.

Computation and Calculations

In this text, the term *computation* refers to the mathematical manipulation of numbers while the term *calculations* refers to the overall presentation including not only the computations but also the statements of explanation that go along with the actual computations.

In addition to being accurate, engineering calculations should be prepared on either special forms created for the particular type of calculation used or on engineering calculation paper with each sheet showing the name of the project, date and the names of preparer and checker.

Also included in the calculations should be all design assumptions made, identification of the computation methodology and computer software used, and a clear depiction of the results, including diagrams if necessary. In general, the calculations should include all information needed for another engineer not familiar with the project to understand the design process and the results.

1.7 Metrication

Traditionally, hydraulic and hydrologic design calculations have been performed in the United States using the English system of units. These include feet, pounds and seconds as base units with a variety of derived units used as well. However, the use of English units currently is evolving into the use of *metric* units also known as the International System of Units (SI). These include meters, kilograms and seconds as base units.

Because most of the world is geared to SI units, the U.S. government decided in 1975 to convert to that system as well. The resulting campaign to "go metric" soon lost momentum however, and the conversion was not completed. Currently, the campaign is in progress again with initiative from the federal government.

Design work performed for a federal agency must in some cases be presented in SI units in order to be approved. However, designs for other agencies such as state, county and local, generally are not required to use SI units. Thus, the overall picture in civil engineering regarding metrication is mixed and somewhat confusing at the present time.

For example, in Canada, which is officially an SI country, civil engineers perform some designs in SI units and some in English depending upon a variety of mitigating factors. Initiative in Canada to become completely metricated flows from the central government.

In the United States, despite the legacy of tenacious adherence to old unit systems, it appears that a complete change to SI units in the near to medium future is on the way. Meanwhile, everyday hydraulic/hydrologic design such as storm sewers, culverts and detention basins continues in English units.

Presentation of design theory and examples in this text is done, to the extent practical, using both systems. Equations which are empirical in nature are presented in both English and SI forms. Rationally derived equations which are valid for any consistent system of units are presented in one form followed by an explanation of appropriate English and SI units. English units are shown first followed by SI units in parentheses. For example, V = volume, ft³ (m³). Unit conversions are presented in Appendix H for the reader's convenience.

Example 1-5

Problem
Find the circumference (in meters) of a pipe having a diameter measured to be 4.00 feet.

Solution
First, convert the diameter to meters.

$$4.00 \text{ feet} \times \frac{1 \text{ m}}{3.28 \text{ ft}} = 1.2195 \text{ m}$$

Next, find the circumference

c = πd
c = (π) (1.2195)
c = 3.83 m   (Answer)

Three significant figures are used in the final answer in accordance with the rules for significant figures.

PROBLEMS

1. Find the cross-sectional area of a pipe with diameter measuring 3.04 feet. Express the answer with the proper significant figures.

2. Compute the volume of pavement in a road measured to be 22.0 feet wide and 1.0 mile long. The measured pavement thickness is 0.650 feet. Express the answer in cubic yards using the proper significant figures.

3. If the depth of water in a stream is measured as $6^1/_2$ inches, find the depth in feet using the proper significant figures.

4. Find the volume of a cylinder with diameter measuring 1.30 feet and length measuring 60. feet. Express the answer in cubic feet using the proper significant figures.

5. Three different lengths of new curb are measured by three different inspectors as follows: 12.25 feet, 151. feet and 25.0 feet. Find the total length of new curb using proper significant figures.

6. Find the cross-sectional area of a pipe with diameter measuring 36 inches. Express the answer in metric units using the proper significant figures.

7. A rectangular field has dimensions 45.00 feet by 125.00 feet. Find the area using the proper significant figures expressed as

   a) square feet   b) acres   c) square meters   d) hectares

8. A roadway is to be designed to carry normal vehicular traffic between two existing parallel roadways separated by a distance of approximately 250 feet. Using your imagination, outline as many factors as possible to be considered in the design.

9. A pedestrian walkway (sidewalk) is to be installed along one side of an existing paved street. Outline as many factors as possible to be considered in the design.

10. A culvert (pipe) is to be designed to convey a small stream through a new road embankment. Outline as many factors as possible to be considered in the design.

FURTHER READING

American Society for Testing and Materials (1993). *Standard Practice for Use of the International System of Units (SI)*. Philadelphia, PA: ASTM.

de Camp, L. (1963). *The Ancient Engineers*. New York: Dorset Press.

Schodek, D. (1987). *Landmarks in American Civil Engineering*. Cambridge, Mass.: MIT Press.

2

FUNDAMENTAL HYDRAULICS

Hydraulics is the study of the mechanics of water at rest and in motion. (The Greek word *hydor* means water.) Since water is a fluid, hydraulics is a branch of fluid mechanics which is a broader study of the behavior of fluids.

In this chapter, we will learn about the fundamental properties of water and how to describe mathematically the flow of water from a higher location to a lower location under the influence of gravity.

OBJECTIVES

After completing this chapter, the reader should be able to:
- Describe certain properties of water, including pressure, viscosity, cohesion, adhesion and buoyancy
- Compute the flow in a simple hydraulic system using Benoulli's Equation
- Sketch the energy grade line and hydraulic grade line for a simple hydraulic system
- Calculate flow through an orifice
- Calculate flow over a weir

2.1   Hydrostatics

The term *hydrostatics* refers to the study of water at rest. Before we describe water in motion it is helpful to discuss some fundamental principles of hydrostatics.

Properties of Water

Even though water is a fluid, it is a material with weight and density just like a solid. And like a solid, water is virtually incompressible. However, the difference between water and solids is all too obvious—water cannot hold its shape without a container and therefore freely flows under the influence of gravity.

However, some solids (those which are particularly plastic) can actually flow under the proper circumstances. So, what is the true distinction between fluids and solids? The answer lies in the nature of the forces between the molecules comprising the substance.

In fluids, molecular forces are such that the substance will yield to the slightest stress while solids, no matter how plastic, require a certain magnitude of stress to be exerted before they will yield. When the molecules of a solid are displaced under external stress, the intermolecular forces tend to restore the molecules to their original positions. But when the molecules of a fluid are so displaced, there is no such restoring force.

Therefore, when a fluid is subjected to an external stress, its molecules readily yield and slide past one another resulting in shearing action. However, one fluid will resist shear stress more than another giving rise to the property of fluids called *viscosity*. Viscosity can be described as a fluid's resistance to shear stress.

Viscosity is sometimes confused with density but it is very different. Whereas *density* refers simply to the amount of mass per unit volume, *viscosity* refers to the ability of the fluid molecules to flow past each other. Thus, a very dense fluid could have a low viscosity or vice versa.

The properties of viscosity and density are well illustrated by the example of oil and water. Most oil, as we know, is less dense than water and therefore floats on water's surface. Yet despite its lack of density, oil is more viscous than water.

The property called viscosity has no practical meaning unless the fluid is moving under the influence of an external stress. In the case of water, the external stress is gravity. *Absolute viscosity* is defined as a measure of the influence of the motion of one layer of a fluid upon another layer a short distance away. Absolute viscosity, designated μ, has units of lb-s/ft^2 (kg-s/m^2). Because it has been found that in many hydraulic problems density is a factor, another form of viscosity, called *kinematic viscosity* has been defined as absolute viscosity divided by density. Kinematic viscosity is designated by υ and has units of ft^2/s (m^2/s). Kinematic viscosity of water is taken as 1×10^{-5} ft^2/s (9.29×10^{-7} m^2/s).

Other properties of water include *cohesion* and *adhesion* which are further descriptions of molecular attraction. Cohesion enables water to resist a slight tensile stress while adhesion enables it to adhere to another body. Figure 2-1 shows some familiar patterns of water in a container caused by cohesion and adhesion. In Figure 2-1(a), adhesion causes the water in the test tube to wet the side for a short distance above the surface. Adhesion results in capillarity in a container of much less diameter. Figure 2-1(b) shows a meniscus at the top of a container caused by surface tension which results from cohesion of molecules at the surface.

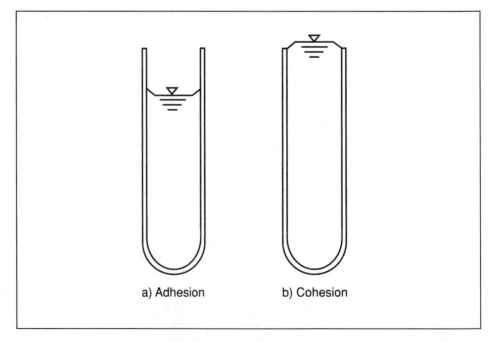

a) Adhesion   b) Cohesion

Figure 2-1 Examples of adhesion and cohesion in water in a glass test tube.

Pressure

The fact that water, like all liquids, has weight but cannot hold its shape gives rise to the unique relation between depth and *pressure*. Pressure is defined as force per unit area, or

$$p = \frac{F}{a} \tag{2-1}$$

where p = pressure, pounds/ft² (N/m²)
       F = force, pounds (Newtons)
       a = area, ft² (m²)

Water in a container exerts pressure at all points on the walls of the container and the magnitude of the pressure varies with the depth. Figure 2-2 shows a schematic description of water pressure contrasted with pressure exerted by a solid. Notice that water pressure is exerted perpendicular, or normal, to the wall surfaces.

To compute the pressure at any point in a vessel of water we can use Equation 2-1 but we must find a way to evaluate F, the force. The force is simply equal to the weight of the vertical column of water above the point.

So in Figure 2-3, the pressure exerted in all directions by the small cube of water is equal to the weight of the column of water divided by the area over which the weight acts, a. The weight is given as

$$\text{Weight} = \gamma V$$

where γ = specific weight, lb/ft³ (N/m³)
       V = volume, ft³ (m³)

The specific weight of water is taken as 62.4 lb/ft³ (9.8 x 10³ N/m³). The volume, V, is equal to the area of the base times the height, or V = az where z is the vertical distance from the water surface to the small cube expressed in ft(m). Therefore, the pressure can be written as

$$p = \gamma \frac{az}{a} = \gamma z. \tag{2-2}$$

a) Water   a) Brick

Figure 2-2   Water pressure is exerted on all walls of its container whereas a solid exerts pressure only downward.

Figure 2-3   Pressure at any point in a volume of water.

Figure 2-4   The pressure at point P is the same in each container provided the vertical depth, z, is the same.

Therefore, water pressure at a depth, z, below a free surface is computed from Equation 2-2. Notice that the area of the small cube of water need not be determined.

It is also important to note that the pressure at any given depth depends only on the depth and not the volume or shape of the container. Thus, the pressures are equal in the three examples in Figure 2-4.

In Figure 2-4(c), the pressure at point P is equal to γz even though there is no vertical column of water from P to the surface. This is because at every depth, pressure is transmitted equally through a horizontal plane resulting in a series of imaginary planes in which pressure is equal throughout any one plane but increases from plane to plane as depth increases. Thus, the pressure at point P is the same *as though* a vertical column of water with height z rested upon it.

Example 2-1

Problem

A reservoir of water is connected to a pipe as shown in Figure 2-5. The pipe valve is closed preventing any flow. Find the water pressure in the center of the pipe at the valve. The elevation of the reservoir surface is 550.0 feet (NGVD) and that of the center of the valve is 525.0 feet (NGVD).

Note: All elevations used in this text are in the National Geodetic Vertical Datum (NGVD) which is given in feet.

Figure 2-5   Reservoir connected to a pipe with valve closed.

Solution
The vertical depth from the surface to the point in question is the difference in elevations.

z = 550.0 – 525.0
z = 25.0 ft.

Therefore, using Equation 2-2

p = γz
p = (62.4) (25.0)
p = 1560 lb/ft^2 (Answer).

The pressure that has been discussed above is called *gauge* pressure, meaning pressure in excess of atmospheric pressure. At the surface of water exposed to the atmosphere (called a *free surface*) the actual pressure is equal to that exerted by the atmosphere and increases with depth below the surface. However, by convention, gauge pressure assumes zero magnitude at the surface. All references to pressure in this text will be to gauge pressure as defined above.

Many instruments have been devised to measure water pressure, but perhaps the one of most interest is the piezometer tube. A *piezometer* is a simple device for measuring pressure, both static and dynamic. A tube with a top exposed to the atmosphere is tapped into the container at the point the pressure is desired. Water in the container immediately rises to a level directly proportional to the pressure. Figure 2-6 shows a schematic diagram of a piezometer.

Figure 2-6   A piezometer indicating pressure in a pipe as the height of water in a vertical tube.

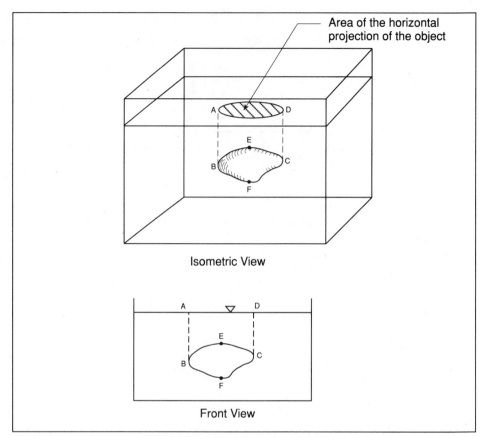

Figure 2-7   Buoyant force on a submerged object.

Buoyancy

Buoyancy is the uplifting force exerted by water on a submerged solid object. Common experience tells us that a heavy rock seems lighter when held under water. This is due to the buoyancy of water.

To understand buoyancy, consider the object shown in Figure 2-7. Essentially, three vertical forces act upon the object: pressure downward by the water above the object, pressure upward by the water below the object and gravity, or the weight of the object. On the upper surface of the body, the vertical component of the pressure is equal to the weight of the volume of water above the object (volume ABECD in Figure 2-7) divided by the horizontal projection of the area of the object. On the lower surface, the vertical component of pressure is equal to the weight of the volume of water above the bottom surface of the object (volume ABFCD in Figure 2-7) divided by the horizontal projection of the area of the object. This follows the same reasoning used to derive Equation 2-2. On the lower surface, the vertical component of pressure is equal to the weight of the volume of water above the bottom surface of the object (volume ABFCD in Figure 2-7) divided by the area of the horizontal projection of the object. This is true even though part of this volume is not occupied by water. It follows the same reasoning used for Figure 2-4(c) to show that a point need not have an actual column of water above it to have a pressure corresponding to the actual column of water.

Since the volume of water above the bottom surface is greater than the volume of water above the top surface, the force pushing up on the object is greater than the force pushing down on the object. The buoyant force is the difference between these two forces and is equal to the weight of water displaced by the object.

If the buoyant force is greater than the weight of the object, the object will rise to the surface of the water. If the buoyant force is less than the weight of the object, the object will sink to the bottom.

Example 2-2

Problem

A 50-pound plastic ball with diameter 2.0 feet is placed in water. What is the buoyant force acting on the ball? Will the ball float or sink?

Solution

The buoyant force is equal to the weight of water displaced by the ball. The volume is

$$V = \frac{4}{3}\pi r^3$$

$$V = \frac{4}{3}\pi(1)^3$$

$$V = 4.19 \text{ ft}^3.$$

The weight of water displaced is

$$W = \gamma V$$
$$W = (62.4)(4.19)$$
$$W = 261 \text{ lb.}$$

Therefore, the buoyant force on the ball is 261 pounds. Since the weight of the ball is less than the buoyant force, the ball will float.   (Answer)

2.2   Hydrodynamics

In this section we will see how to describe mathematically the flow of water from a higher location to a lower location under the influence of gravity. Examples of such flow include those shown in Figure 2-8.

a) Reservoir discharing
   into a pipe

b) Orifice flow

c) Reservoir discharging through
   a pipe to a lower reservoir

Figure 2-8   Examples of gravity flow of water.

Steady Flow

The applications of hydrodynamics we will discuss involve *steady flow* in which the rate of flow does not vary significantly with respect to time at any point along the stream. For example, a water faucet which is left running for a period of time exhibits steady flow.

By contrast, *unsteady flow* involves the change of the water's flow rate at any point in the stream. Thus, as you turn on a faucet you create unsteady flow in the pipe. Another example of unsteady flow would be the quickly increasing flow in a channel just down-stream of a dam which has just breached or broken open.

The phenomenon called *water hammer* is the extreme variations in pressure within a pipe caused by an abrupt stoppage in flow. So if, for example, a valve is quickly closed, the pressure against the valve surges due to the collision of the moving water with the closed valve, sending fluctuations of pressure through the pipe due to the conservation of momentum.

Even the steadiest flow of water has some fluctuation associated with it. However, for practical purposes, steady flow is that in which the fluctuations are minor and tend to average to zero over time.

Conservation Laws

When water flows from a higher elevation to a lower elevation, it follows the laws of physics just as any other object would. Some of those laws are expressed as conservation laws such as conservation of energy, momentum and mass.

We will consider conservation of energy and of mass in our study of hydrodynamics. Conservation of momentum is used to describe hydraulic phenomena beyond the scope of this book and will not be discussed.

Conservation of energy is very important in describing the behavior of water under-going steady flow. Neglecting friction, potential energy at the higher elevation gives way to kinetic energy at the lower elevation resulting in the following equation familiar from our study of physics

$$U_1 + K_1 = U_2 + K_2 \qquad (2\text{-}3)$$

where U_1 and K_1 represent potential and kinetic energies respectively at point 1, and U_2 and K_2 represent potential and kinetic energies respectively at point 2.

However, what makes water different from the familiar objects studied in physics is water's inability to maintain a constant shape. Therefore, the energy equation must take on a new form to account for water's fluid nature.

To do this, we represent the energy per unit weight of water instead of just energy alone. Thus, Equation 2-3 becomes

$$\frac{U_1}{mg} + \frac{K_1}{mg} = \frac{U_2}{mg} + \frac{K_2}{mg} \qquad (2\text{-}4)$$

where mg represents the weight of a small elemental volume of water as shown in Figure 2-9 as it moves through a hydraulic system. Equations 2-3 and 2-4 are consistent with both English and SI units.

Figure 2-9 Profile of flow from a reservoir through a pipe.

Station 1 and Station 2 are points along the profile of the hydraulic system shown in Figure 2-9. Mathematically, the system is taken as one-dimensional; that is, the energy is considered at any station along the profile as if the system flows in a straight line.

At Station 1, the elemental volume of water is at some depth, z, below the surface and at some height, h, above an arbitrary datum. The potential energy U_1 is defined as

$$U_1 = mgh_1 + mgz_1$$

where the second term, mgz_1, is due to the pressure exerted on the elemental volume by the water above it. Potential energy per unit weight is

$$\frac{U_1}{mg} = \frac{mgh_1}{mg} + \frac{mgz_1}{mg} = h_1 + z_1.$$

However, as we saw earlier in Equation 2-2, z may be expressed as p/γ. Therefore, potential energy per unit weight at Station 1 may be written as

$$\frac{U_1}{mg} = h_1 + \frac{p_1}{\gamma}.$$

You might notice that U_1 will be the same value for any elemental volume of water with mass m along the vertical line comprising Station 1. In fact, if the elemental volume is located at the surface of the water, $p_1 = 0$ and $U_1/mg = h_1$.

The kinetic energy per unit weight at Station 1 is

$$\frac{K_1}{mg} = \frac{\frac{1}{2}mv_1^2}{mg} = \frac{v_1^2}{2g}$$

where v_1^2 is the velocity of the water element.

At Station 2, the water is contained within a pipe and has no free surface, so for ease of mathematics we consider the water element at the center of the pipe. The potential energy per unit weight at Station 2 is

$$\frac{U_2}{mg} = \frac{mgh_2}{mg} + \frac{p_2}{\gamma}$$

where h_2 is the height of the center of the pipe above the datum and p_2 is the pressure exerted by the reservoir above the point and γ is the specific weight of water.

The pressure term, p_2/γ, can be thought of as the potential energy contained in each pound (Newton) of water in the form of pressure. That is, the pressure at Station 2 has the ability to do p_2/γ foot-pounds (N-m) of work on every pound of water.

The kinetic energy per unit weight at Station 2 is

$$\frac{K_2}{mg} = \frac{\frac{1}{2}mv_2^2}{mg} = \frac{v_2^2}{2g}$$

where v_2 is the velocity of the water element at Station 2.

Now Equation 2-4 can be rewritten as

$$h_1 + \frac{p_1}{\gamma} + \frac{v_1^2}{2g} = h_2 + \frac{p_2}{\gamma} + \frac{v_2^2}{2g}.$$

(2-5)

Equation 2-5 is a form of the Bernoulli Equation for an ideal condition in which friction is not present. Equation 2-5 is consistent with both English and SI units.

In cases involving a free surface at Station 1, Equation 2-5 is simplified to

$$h_1 + \frac{v_1^2}{2g} = h_2 + \frac{p_2}{\gamma} + \frac{v_2^2}{2g}$$

(2-6)

where h_1 represents the vertical distance from the arbitrary datum to the free surface and $p_1 = 0$. Equation 2-6 also is consistent with both English and SI units.

Example 2-3

Problem
Determine the velocity of water at the outflow end of the pipe shown in Figure 2-9 neglecting friction if the elevation of the center of the pipe is 525 feet (NGVD), and the surface of the reservoir is 550 feet (NGVD).

Solution
Flow in the reservoir has such low velocity that it may be considered to be zero. Therefore, $v_1 = 0$ ft/s. Also, since the outflow has a free surface, the pressure term becomes zero. Substituting into Equation 2-6 then gives

$$550 + 0 = 525 + \frac{v_2^2}{(2)\,(32.2)}$$

$$v_2^2 = (550 - 525)\,(2)\,(32.2)$$

$$v_2^2 = 1610$$

$$v_2 = \sqrt{1610} = 40.1\,\text{ft/s}. \quad \text{(Answer)}$$

An interesting factor to notice in this example is that the velocity at point 2 does not depend on the size of the reservoir, only on its height. Likewise, the velocity does not depend on the slope of the pipe or the diameter of the pipe. Difference in elevation is the only determining factor.

When the energy of water is described as energy per unit weight, its dimension becomes feet (meters), and it is called *head*. Thus, the total energy of the water at Station 1 in Figure 2-9 is described as its total head which is the number of feet above an arbitrary datum. At Station 2, the pressure component of potential energy is described as pressure head and kinetic energy is described as velocity head and both terms have the units of *feet (meters)* as do all terms in Bernoulli's Equation. Since all energy terms are measured in feet (meters), a graphical representation of energy becomes convenient for hydraulic systems. Thus, the system described in Figure 2-9 can be re-drawn as shown in Figure 2-10 to show two important lines used to analyze hydraulic systems. The *energy grade line* depicts total energy or total head at all stations along the system. In this case, the energy grade line is horizontal since total energy remains constant (no energy lost to friction). The *hydraulic grade line* depicts potential energy (position plus pressure head) at all stations along the system. The vertical separation between the energy and hydraulic grade lines is the velocity head. Wherever there is a free surface, the hydraulic grade line is coincident with it.

A close scrutiny of Figure 2-10 reveals that the hydraulic grade line is coincident with the free water surfaces at both ends. The energy grade line is coincident only with the free surface of the reservoir because as we have seen previously the reservoir velocity is negligible.

To better describe the motion of water in the real world, however, friction must be accounted for. Since energy is lost from the system due to friction, the total energy, or head, at Station 2 will be less than the total energy, or head, at Station 1 by an amount equal to the friction loss. Thus, Bernoulli's Equation takes on the form

$$h_1 + \frac{v_1^2}{2g} = h_2 + \frac{p_2}{\gamma} + \frac{v_2^2}{2g} + h_L \tag{2-7}$$

where h_L represents the amount of head lost to friction. Equation 2-7 therefore depicts a more realistic form of Bernoulli's Equation for a hydraulic system with a free surface at Station 1. Figure 2-11 represents the hydraulic system shown in Figure 2-10, except with friction accounted for.

Figure 2-10 Energy grade line and hydraulic grade line for an ideal hydraulic system with no friction.

Figure 2-11   Energy and hydraulic grade lines for a hydraulic system in which friction losses are included.

A close scrutiny of Figure 2-11 reveals that energy losses take place in at least two ways. First, there is a small but sudden drop of the energy grade line at the point where water enters the pipe from the reservoir. This drop, called *entrance loss*, is due to a loss of energy caused by the turbulent motion of the water as it enters from the larger reservoir to the more restrictive pipe. Secondly, a constant drop in energy takes place along the entire length of the pipe. This drop is due to disruption of the flow caused by contact with the inside surface of the pipe and is called *friction loss*. Total head loss, h_L, is the sum of entrance loss h_e and friction loss h_f. Thus, $h_L = h_e + h_f$.

Friction in hydraulics is different from the friction associated with solid objects sliding over each other. When water slides over a surface, particles in contact with the surface are slowed down, creating a velocity gradient, or unequal distribution of velocity. Figure 2-12 shows a typical distribution or gradient in pipe flow with the slowest velocity at the edges and the greatest velocity in the center. (The term v used in Equation 2-5 is the average velocity of all the velocities in the distribution.)

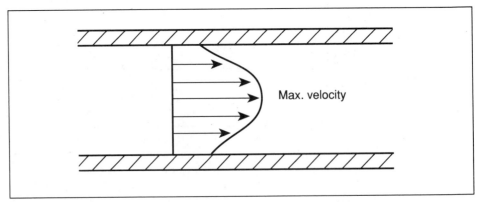

Figure 2-12   Velocity distribution of water flowing in a pipe.

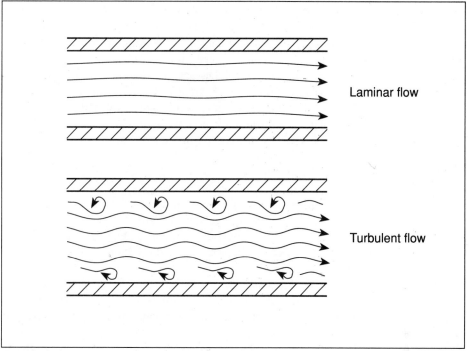

Figure 2-13 Laminar flow and turbulent flow.

The amount of friction loss depends upon the velocity of flow as well as the roughness of the pipe. When velocity is relatively low, the flow is smooth and is called *laminar* flow. This type of flow experiences very little friction loss as it traverses the conduit containing it. This is because the layers of flow do not move relative to each other. As velocity increases, the velocity gradient within the mass of moving water increases until a point is reached where eddies develop and the flow becomes rough. This type of flow is called *turbulent* and results in much more friction loss. Most of the flow encountered in hydraulic engineering is turbulent flow. Figure 2-13 illustrates laminar and turbulent flows.

The dimensionless parameter called the Reynolds Number, N_R, is used to describe the degree of turbulence in flowing water:

$$N_R = \frac{Dv}{\upsilon} \tag{2-8}$$

where D = diameter of pipe, ft (m),
      v = average velocity, ft/s (m/s),
      υ = viscosity of water, 1×10^{-5} ft²/s (9.29×10^{-7} m²/s).

Low values of N_R (up to 2000) describe smooth or laminar water flow while large values of N_R (above 10,000) indicate turbulent flow.

The expression for friction loss for turbulent flow in circular pipes is given by

$$h_f = f \frac{Lv^2}{2gD} \tag{2-9}$$

where h_f = friction head loss, ft (m),
     f = friction factor,
     L = length of pipe, ft (m),
     D = diameter of pipe, ft (m),
     v = average velocity, ft/s (m/s),
     g = acceleration of gravity, 32.2 ft/s² (9.81 m/s²).

The dimensionless friction factor, f, is an empirically derived parameter depending upon a complex set of flow conditions. Figure 2-14 depicts the Moody Diagram which is a graphical solution for f depending upon such values as viscosity, υ, and roughness, ε, and Reynolds Number, N_R.

In addition to conservation of energy, flow of water is further described by conservation of mass which, simply stated, asserts that mass is neither created nor destroyed.

This means that for an incompressible fluid such as water flowing in a stream or pipe, the quantity of mass passing a cross section at Station 1 per unit time is equal to the quantity of mass passing a cross section at Station 2, or

$$a_1 v_1 = a_2 v_2 \qquad (2\text{-}10)$$

where a_1 = cross sectional area at Station 1, ft^2(m^2),
  v_1 = average velocity at Station 1, ft/s (m/s),
  a_2 = cross sectional area at Station 2, ft^2(m^2),
  v_2 = average velocity at Station 2, ft/s (m/s).

Equation 2-10 is called the *continuity equation* and applies to water flowing in any conduit (pipe, channel, stream) as long as no water enters or leaves the conduit between Station 1 and Station 2. The continuity equation also is expressed as

$$Q = va \qquad (2\text{-}11)$$

where Q = quantity (or rate) of flow in cubic feet per second, cfs (m^3/s),
  v = average velocity across a given cross section of flow in ft/s (m/s),
  a = area of the cross section in ft^2 (m^2).

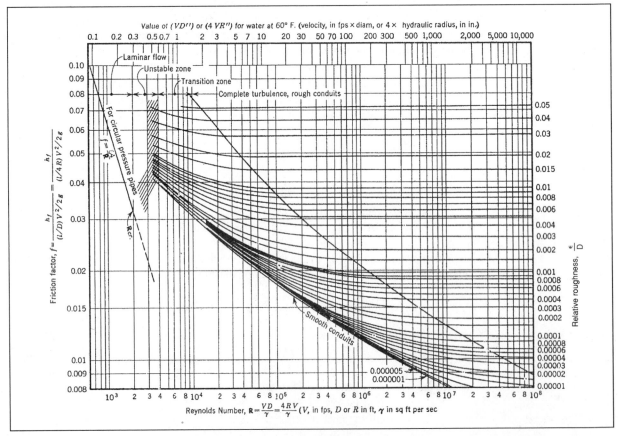

Figure 2-14 Moody Diagram—friction factor for pipes. *(Courtesy of ASCE, Design and Construction of Sanitary and Storm Sewers.)*

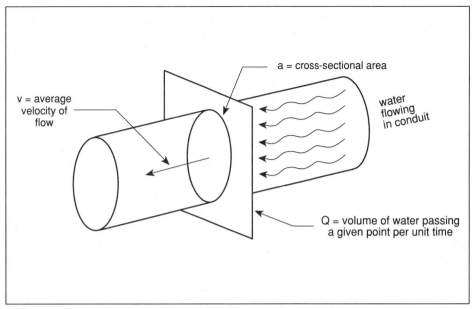

v = average
velocity of
flow

a = cross-sectional area

water
flowing
in conduit

Q = volume of water passing
a given point per unit time

Figure 2-15 Definitions of Q, v and a—parameters used in the continuity equation.

Figure 2-15 illustrates these parameters.

Example 2-4

Problem

Determine the quantity of flow, Q, at the outflow end of the pipe shown in Figure 2-9 if the elevation of the center of the pipe is 525 feet and the surface of the reservoir is 550 feet. The pipe is composed of smooth concrete and has diameter of 2.0 feet and a length of 350 feet.

Solution

As in Example 2-3, $v_1 = p_2/\gamma = 0$. Also, assume $h_e = 0$. Therefore, Equation 2-7 becomes

$$h_1 + 0 = h_2 + 0 + \frac{v_2^2}{2g} + h_f$$

Substituting Equation 2-9 gives

$$h_1 = h_2 + 0 + \frac{v_2^2}{2g} + f\frac{Lv_2^2}{2gD}$$

which can be rearranged to

$$h_1 - h_2 = \{1 + f\frac{L}{D}\}\frac{v_2^2}{2g}. \tag{2-12}$$

Substituting known values gives

$$25 = \{1 + f\frac{350}{2}\}\frac{v_2^2}{64.4}$$

$$1610 = \{1 + 175f\}v_2^2.$$

This equation containing f and v_2 can be solved using an iteration process as follows:

Assume a value for f, say f = 0.02. Using f = 0.02, v_2 = 18.9 ft/s from the above equation. From Equation 2-8, the Reynold's Number is

$$N_R = \frac{(18.9)\ (2)}{10^{-5}}$$

$$N_R = 3.78 \times 10^6.$$

Now, look at the Moody Chart in Figure 2-14 to find the resulting value of f. If the value of f found on the Moody Chart matches the assumed value, then the assumed value was correct and the resulting value of v_2 would be correct as well.

To find f on the Moody Chart, first determine ε/D. From Table 2-1, ε = 0.001. Therefore,

ε/D = 0.0005.

On the Moody Chart, $N_R = 3.78 \times 10^6$ intersects ε/D = 0.0005 at approximately f = 0.018. Since this value of f does not agree with the assumed value of 0.020, choose another f.

Choose f = 0.018.

Then, from Equation 2-12, v_2 = 19.7 ft/s. The resulting Reynold's Number is

$$N_R = \frac{(2)\ (19.7)}{10^{-5}}$$

$$N_R = 3.9 \times 10^6.$$

Table 2-1
Typical values of Roughness, ε

Pipe Material	ε (ft)
Riveted steel, few rivets	0.003
Riveted steel, many rivets	0.030
Concrete, finished surface	0.001
Concrete, rough surface	0.010
Wood-stave, smooth surface	0.0006
Wood-stave, rough surface	0.003
Cast iron, new	0.00085
Galvanized iron, new	0.00050
Drawn tubing, new	0.000005

Courtesy of ASCE, Design and Construction of Sanitary and Storm Sewers.)

On the Moody Chart, $N_R = 3.9 \times 10^6$ intersects $\varepsilon/D = 0.0005$ at approximately $f = 0.018$. Therefore, $v_2 = 19.7$ ft/s is correct. Finally, using Equation 2-11

$$Q = v_2 a$$

$$Q = (19.7) \ (3.14)$$

$$Q = 61.9 \text{ cfs.} \qquad \text{(Answer)}$$

Comparison with Example 2-3 reveals that the velocity at Station 2 was found to be about half the velocity computed assuming no friction. Also, note that even though pipe friction was used in this example, entrance loss was still neglected. In general, entrance loss constitutes a relatively minor portion of the total head loss encountered.

Some other examples of hydraulic systems are shown in Figure 2-16. In Figure 2-16(a), water flows from a higher reservoir to a lower reservoir through a pipe. Notice that the energy and hydraulic grade lines are not affected by the path taken by the pipe. They are determined only by the pipe's roughness, diameter and connections to the reservoirs.

Figure 2-16   Examples of hydraulic systems showing energy grade line and hydraulic grade line.

In Figure 2-16(b), the pipe connecting the reservoirs abruptly increases in size. Notice that the change in pipe size causes a drop in the energy grade line (due to turbulence) and a corresponding rise in hydraulic grade line (due to a drop in pressure caused by the decreased velocity). The fact that the energy and hydraulic grade lines become closer together indicates the decreased velocity that could have been predicted using the continuity equation. Notice also that if the pipe was drawn horizontally, the energy and hydraulic grade lines would be virtually unchanged.

In Figure 2-16(c), a very different situation is shown. In this case, water flows in a pipe only partially full and therefore has a free surface throughout the hydraulic system. For this reason, the hydraulic grade line is coincident with the water surface. At the end of the pipe, the water spreads onto the ground and looses velocity as shown by the closer hydraulic and energy grade lines.

2.3   Orifice Flow

An *orifice* is a hole in the wall of a container through which water may flow. Generally, the wall is assumed to be thin and the edges of the hole square but many variations of geometry are possible. Figure 2-17 shows examples of orifices.

The amount of flow, Q, passing through the orifice depends upon the pressure at the orifice depth which depends upon the head, h. The flow is also affected by the loss of energy due to turbulence as the water enters the orifice. Figure 2-18 shows how head is defined for two conditions of orifice flow. Flow through an orifice is computed using the orifice equation

$$Q = ca\sqrt{2gh} \qquad\qquad (2-13)$$

where   Q = rate of flow, cfs (m³/s),
         c = entrance coefficient,
        a = cross sectional area of opening, ft² (m²),
        h = total head, ft (m).

a) Thin wall, square edge   b) Thick wall, square edge   c) Thick wall, bevelled edge

Figure 2-17   Examples of orifices.

a) For free discharge, h is measured to the center of the orifice.

b) For submerged discharge, h is measured to the lower water surface.

Figure 2-18   Definition of h for two orifice flow conditions.

The *entrance coefficient*, c, is a dimensionless proportionality constant which accounts for the reduction of flow due to entrance head loss. The experimental value of c for square edged orifices varies depending upon the size and shape of the orifice and the amount of head. However, for most applications reliable results can be achieved using c = 0.62.

Example 2-5

Problem

Find the discharge, Q, through a 6-inch diameter square edged orifice discharging freely. The water surface elevation of the impoundment is 220.0 and the elevation of the center of the orifice is 200.0.

Solution

First, find the cross-sectional area, converting 6 inches to 0.50 feet:

$$a = \pi r^2$$
$$a = \pi(.25)^2$$
$$a = 0.196 \text{ ft}^2.$$

Next, find total head as the difference between elevations:

$$h = 220.0 - 200.0$$
$$h = 20.0 \text{ ft.}$$

Therefore, using Equation 2-13,

$$Q = ca\sqrt{2gh}$$

$$Q = (.62)(.196)\sqrt{(2)(32.2)(20)}$$

$$Q = 4.4 \text{ cfs.} \qquad \text{(Answer)}$$

Example 2-6

Problem

Find the discharge, Q, through a 4" x 6" rectangular orifice with square edges discharging submerged. The water surface elevation of the impoundment is 220.0 and the elevation of the water surface of the discharge water is 215.0. The elevation of the center of the orifice is 200.0.

Solution

First, find the cross-sectional area in square feet:

$$a = \left[\frac{4}{12}\right]\left[\frac{6}{12}\right]$$

$a = 0.167$ ft^2.

Next, find total head as the difference between elevations:

$h = 220.0 - 215.0$

$h = 5.0$ ft.

Note that the elevation of the orifice is not relevant. Therefore, using Equation 2-13,

$$Q = ca\sqrt{2gh}$$

$$Q = (.62)(.167)\sqrt{(2)(32.2)(5)}$$

$Q = 1.9$ cfs (Answer)

2.4 Weir Flow

A *weir* is a structure which, like an orifice, regulates the flow of water out of an impoundment or reservoir. Generally, a weir consists of a horizontal surface over which water is allowed to flow. Typical uses include outlet structures for dams and detention basins as well as other impoundments such as holding tanks in sewage treatment plants. Also, weirs are widely used as measuring devices with such applications as natural streams and treatment works.

Figure 2-19 shows a typical weir discharging freely to a downstream channel. The energy which pushes water over the crest is measured by the head H above the crest.

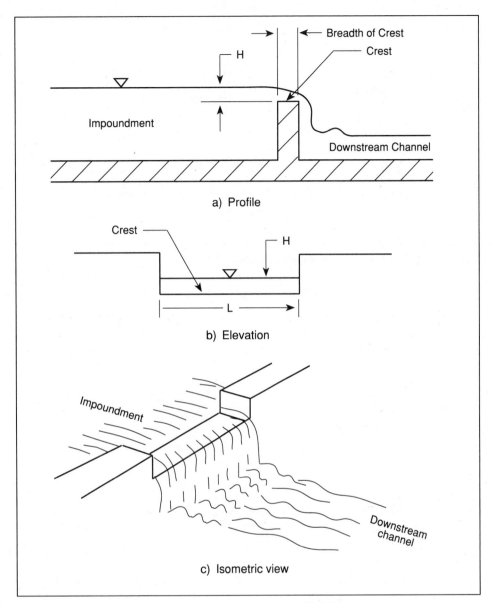

Figure 2-19   Typical Weir.

Close inspection of Figure 2-19(a) reveals that the surface of the impoundment begins to drop as it approaches the weir. This is due to the increase of velocity which is compensated for by a drop in cross sectional area in accordance with the continuity equation. Therefore, H must be measured at some distance away from the weir crest where the velocity is virtually zero. Usually, this location is at least a distance 2.5H upstream of the crest.

Many different types of weirs have been devised over the years by varying the width and shape of the crest. The most important of these are shown in Figures 2-20 and 2-21.

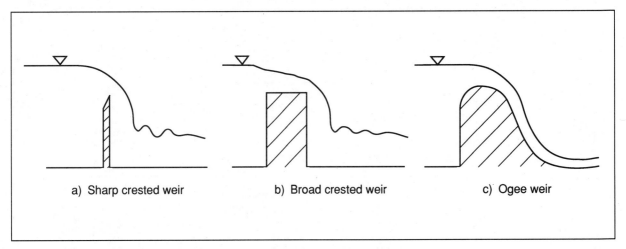

a) Sharp crested weir   b) Broad crested weir   c) Ogee weir

Figure 2-20   Classification of weirs by crest profile.

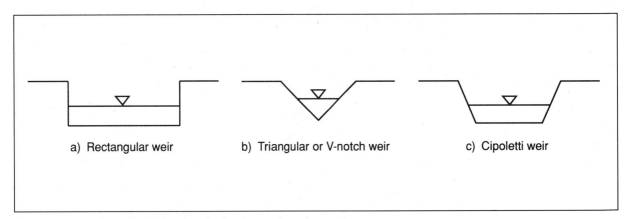

a) Rectangular weir   b) Triangular or V-notch weir   c) Cipoletti weir

Figure 2-21   Classification of weirs by crest elevation.

Rectangular Weir

The rate of flow or discharge over a rectangular weir is computed by the weir formula

$$Q = cLH^{3/2} \qquad\qquad (2\text{-}14)$$

where Q = discharge, cfs,
      c = discharge coefficient,
      L = effective crest length, ft,
      H = head above crest, ft.

Equation 2-14 is used for most rectangular weir flow computations using the English system of units. The *discharge coefficient* is an empirically determined multiplier which accounts for a number of hydraulic factors which would be difficult to describe mathematically. Discharge coefficients depend upon the type of weir and the depth of flow. Selected values are listed in Appendix A-5 for use in solving weir flow problems. To convert Equation 2-14 for use with the metric system of units, appropriate values of c would be required. Values of c in Appendix A-5 are intended for use with English system of units.

Figure 2-22   Hydraulic elements of a sharp-crested weir.

Sharp-Crested Weir

A rectangular, sharp-crested weir is perhaps the most basic type and is used for measuring flow in a channel and also as a simple spillway structure. Computation of the discharge, Q, for this type of weir depends upon the dimensions of the weir in relation to the channel and also to the head, H. Figure 2-22 illustrates the parameters used in determining Q for a sharp crested weir.

As water flows past the vertical sides of the weir, a loss of energy takes place which is called *contraction*. This contraction can be accounted for by reducing the actual measured length of the weir to a lesser value called *effective length*. Effective length, L, is computed by

$$L = L' - 0.1nH \tag{2-15}$$

where L' = actual measured crest length, ft,
  n = number of contractions,
  H = head above the crest, ft.

If the weir is centered in the channel with L' less than channel width, B, there are two end contractions and $n = 2$. If $L' = B$, there are no end contractions and $n = 0$. If the weir crest is against one side of the channel and not the other, there is one contraction and $n = 1$.

The height of the weir above the channel bottom also has an effect on the discharge, Q. This effect is accounted for by adjusting the discharge coefficient, c, in accordance with the height, P. Thus

$$c = 3.27 + 0.40 \frac{H}{P} \qquad (2\text{-}16)$$

where P = height of crest above the channel bottom, ft.

Example 2-7

Problem
Find the discharge flowing over a sharp crested weir as shown below if the crest elevation is 282.00 and the impoundment elevation is 283.75.

Solution
First, find H from the elevations given

$$H = 283.75 - 282.00$$

$$H = 1.75 \text{ ft.}$$

Since the weir has two contractions, n = 2. Therefore, using Equation 2-15

$$L = L' - 0.1nH$$

$$L = 5.33 - (.1)(2)(1.75)$$

$$L = 4.98 \text{ ft.}$$

Next, find c using Equation 2-16

$$c = 3.27 + 0.40 \frac{H}{P}$$

$$c = 3.27 + (.40) \frac{1.75}{2.85}$$

$$c = 3.516.$$

Finally, find Q using Equation 2-14

$$Q = cLH^{3/2}$$

$$Q = (3.516)\,(4.98)\,(1.75)^{3/2}$$

$$Q = 40.5 \text{ cfs.} \qquad \text{(Answer)}$$

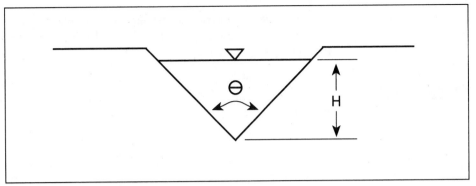

Figure 2-23   Hydraulic elements of a V-notch weir.

V-Notch Weir

A variation of the sharp crested weir is the triangular, or V-notch weir which is used to measure flow when very low quantities are expected. Discharge over a V-notch weir is computed by a variation of Equation 2-14, namely

$$Q = c\{\tan\frac{\theta}{2}\}H^{5/2} \qquad\qquad (2\text{-}17)$$

where θ is the angle (in degrees) made by the notch as shown in Figure 2-23. Although c varies under different conditions, it is taken generally to be 2.5.

Cipoletti Weir

A *Cipoletti* weir is a trapezoidal variation of the sharp- crested weir devised to compensate for loss of flow quantity due to contractions at the vertical edges of a rectangular weir. By sloping the edges at approximately 1:4 (horizontal to vertical) the increasing cross-sectional area of flow, as H increases, compensates for loss of flow due to end contraction.

Broad-Crested Weir

A *broad-crested*, rectangular weir is commonly employed in outlet structures for dams and detention basins. Discharge is computed using the weir equation, Equation 2-14, with values of c found in Appendix A-5. Correction for side contractions generally is not required for this type of weir. A variation of the broad-crested weir used to more precisely regulate discharge is the multi-stage weir as shown in Figure 2-24.

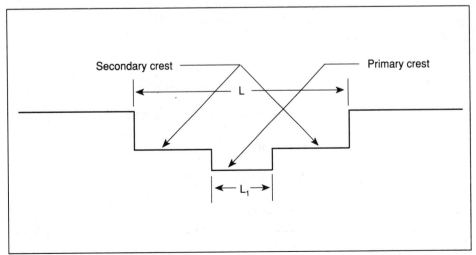

Figure 2-24   Elevation view of two-stage weir.

Discharge for the two-stage weir shown in Figure 2-24 is computed by adding the discharge for the primary crest and the discharge for the secondary crest as detailed in Example 2-10.

Example 2-8

Problem
Find the discharge flowing over a 60° V-notch weir if H = 4.0 inches.

Solution
First, find θ/2:

$$\frac{\theta}{2} = \frac{60}{2} = 30°.$$

Next, find H in feet:

H = 4.0" = 0.33 ft.

Next, c is assumed to be:

c = 2.5.

Finally, find Q from Equation 2-17:

$$Q = c\{\tan\frac{\theta}{2}\}H^{5/2}$$

$$Q = (2.5)\ (\tan 30°)\ (.33)^{5/2}$$

Q = 0.090 cfs.   (Answer)

Example 2-9

Problem
Find the discharge flowing over a rectangular broad-crested weir with a breadth 9.0 inches, length 4.0 feet and a head of 1.25 feet.

Solution
First, find c using the chart in Appendix A-5:

c = 3.215.

(The value of c is found by interpolating between values for H = 1.2 and H = 1.4.)

Then, compute Q using Equation 2-14:

$$Q = cLH^{3/2}$$

$$Q = (3.215)\ (4.0)\ (1.25)^{3/2}$$

Q = 18 cfs.   (Answer)

Example 2-10

Problem

Find the discharge flowing over a two stage weir like that shown in Figure 2-24 if the elevation of the impoundment is 235.50. The overall length of the weir is 22.00 feet and the length of the primary crest is 4.00 feet. The crest breadth is 1.00 foot. The elevation of the primary crest is 233.00 and that of the secondary crest is 233.50.

Solution

Step 1:

Find the discharge, Q_1, over the primary crest. Head, H_1, is found from the elevations of the water surface and the crest:

$$H_1 = 235.50 - 233.00$$

$$H_1 = 2.50 \text{ feet.}$$

Then using the chart in Appendix A-5, find c_1:

$$c_1 = 3.31.$$

Therefore, discharge Q_1 is

$$Q_1 = c_1 L_1 H_1^{3/2}$$

$$Q_1 = (3.31)\ (4.00)\ (2.50)^{3/2}$$

$$Q_1 = 52.3 \text{ cfs.}$$

Step 2:

Find the discharge Q_2 over the secondary crest. First find head, H_2:

$$H_2 = 235.50 - 233.50$$

$$H_2 = 2.00 \text{ ft.}$$

Then, using the chart in Appendix A-5, find c_2:

$$c_2 = 3.30.$$

Therefore, discharge Q_2 is

$$Q_2 = c_2 L_2 H_2^{3/2}$$

$$Q_2 = (3.30)\ (22.0 - 4.00)\ (2.00)^{3/2}$$

$$Q_2 = 168.0 \text{ cfs.}$$

Step 3:

Finally, find total discharge, Q:

$$Q = Q_1 + Q_2$$

$$Q = 52.3 + 168.0$$

$$Q = 220. \text{ cfs.} \qquad \text{(Answer)}$$

Ogee Weir

Another type of rectangular weir commonly used as spillway for a dam is the *ogee weir* shown in Figure 2-20(c). The smooth rounded surface of the oqee weir is designed to reduce energy loss by edge contraction as water passes over the crest, thus increasing the discharge Q for a given head H in comparison to sharp or broad-crested weirs. Computation of discharge is not included in this text since applications of oqee spillways are not considered.

It should be noted that the geometrical shape of an oqee weir is formed by compound curves with radii dependent on a specific anticipated head. However, the term is commonly applied to weirs having the general shape shown in Figure 2-20(c).

PROBLEMS

1. What is the weight in pounds of 100 gallons of water?

2. A swimming pool has depths of 4.0 feet and 12.0 feet at the shallow and deep ends, respectively. Find the pressure in psf at each end.

3. A reservoir of water is connected to a 12-inch diameter pipe 50 feet long and capped at the end. The water surface elevation of the reservoir is 82.5 and the capped end of the pipe is at elevation 38.0. Find the pressure in psf at the capped end of the pipe.

4. A 0.75-pound can of soda with dimensions 2½" x 5" is placed in water.

   (a) What is the buoyant force acting on the can?

   (b) Will the can float or sink?

5. A 2.5-pound brick with dimensions 2½" x 3½" x 8" is placed in water.

   (a) What is the buoyant force acting on the brick?

   (b) Will the brick float or sink?

6. A reservoir of water is connected to an 8-inch diameter pipe 50 feet long and discharging freely. The water surface elevation of the reservoir is 410.0 and the elevation of the center of the pipe at its discharge end is 372.5. Neglecting friction, determine the velocity, v, and discharge, Q, at the end of the pipe.

7. Find the velocity, v, and discharge, Q, for the pipe in Problem 6 if the pipe is composed of new cast iron and friction is considered.

8. Sketch the energy grade line and hydraulic grade line for the hydraulic system shown below.

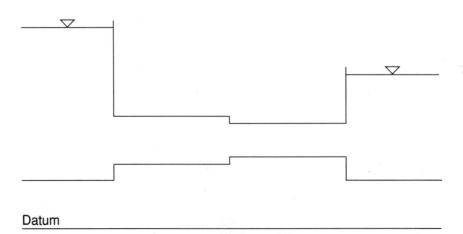

Datum

9. Find the discharge, Q, through a 15-inch diameter orifice in a vertical wall impounding water at a depth of 6.5 feet above the center of the orifice. The orifice has square edges and discharges freely.

10. An artificial lake is impounded by an earth dam with a 12" x 12" orifice for drawdown. The elevation of the surface of the lake is 928.25 and the invert of the orifice is 905.75. When the orifice is opened, what discharge in cfs flows through into the stream below the dam?

11. A reservoir of water is drained by a 10-inch diameter square-edged orifice with center at elevation 289.12 that flows directly into a stream with water surface at elevation 293.49. The elevation of the reservoir surface is 296.85. Find the flow through the orifice.

12. Find the discharge over a 90° V-notch weir if the head is 7.5 inches.

13. Find the discharge over the rectangular sharp-crested weir shown below.

14. Find the discharge over a broad-crested weir with crest breadth of 18 inches and dimensions as shown in Problem 13.

15. Find the discharge over the rectangular, two-stage weir shown below if the breadth of each crest is 2.00 feet.

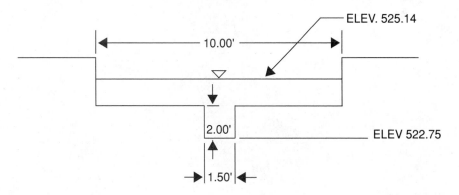

FURTHER READING

Brater, E. F. and King, Horace. (1976). *Handbook of Hydraulics.* New York: McGraw-Hill,

Hita, C. E. and Hwang, Ned H. C. (1987). *Hydraulic Engineering Systems* (2nd ed.). Englewood Cliffs, NJ: Prentice Hall.

Morris, H. M. Wiggert, J. M. (1972). *Applied Hydraulics in Engineering* (2nd ed.). New York: John Wiley & Sons.

Prasuhn, Alan L. (1987). *Fundamentals of Hydraulic Engineering.* New York: Holt, Rinehart and Winston.

3

OPEN CHANNEL HYDRAULICS

When water flows downhill in any conduit with the water surface exposed to the atmosphere (free surface), it is said to undergo *open channel flow*. This type of flow is different from pressure flow in a closed conduit but no less complex. Open channel hydraulics is the study of the mechanics of water flowing in open conduits which generally include channels, streams and even pipes in which the depth is not all the way to the top. An example of open channel flow is shown in Figure 3-1 which depicts a man-made rectangular channel conveying a stream through a park.

OBJECTIVES

After completing this chapter, the reader should be able to:

- Compute the slope of a channel
- Compute normal depth in a channel or pipe
- Compute normal depth in a stream including overbanks
- Compute critical depth in a channel or pipe
- Properly use the channel and pipe design charts
- Analyze a culvert for inlet and outlet control
- Compute headwater depth for a culvert

3.1 Fundamental Concepts

To analyze open channel flow, let us first consider a long uniform channel such as the concrete channel in Figure 3-2(a). A uniform channel is one which maintains a constant shape and slope (or gradient). For all our work with open channel hydraulics, we will assume constant flow rate, Q. *Gradient*, s, is defined as vertical fall divided by horizontal run of a channel or stream and has units of ft/ft (m/m). So, a channel which drops 1.00 feet (m) in a length of 20.0 feet (m) has a gradient of 0.050 ft/ft (m/m). That is,

$$s = \frac{drop}{length}$$

$$s = \frac{1.00}{20.0} = 0.050 \text{ ft/ft (m/m).}$$

Figure 3-1 Open channels are often used to convey stream flow in developed areas. This channel has a rectangular shape with stone sides.

Gradient can also be expressed as a percent by multiplying by 100%. When expressed as a percent, s will be referred to as *slope*. Thus the slope of a channel with s = 0.050 ft/ft (m/m) is 5.0%.

Normal Depth

When water flows in a uniform channel, after an initial transition, it reaches and maintains a constant velocity and constant depth called *normal depth*. Normal depth is defined as the vertical distance from the channel bottom to the water surface. The energy grade line is parallel to the surface because the loss of energy due to friction is exactly compensated for by the acceleration due to gravity.

Various empirical formulas have been devised to compute normal depth in a uniform channel. These formulas bear the names of their creators and include the Chezy, Darcy, Kutter and Manning equations. In the next section, we will look at one of these, the Manning Equation.

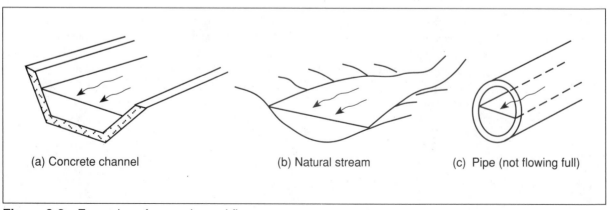

(a) Concrete channel   (b) Natural stream   (c) Pipe (not flowing full)

Figure 3-2 Examples of open channel flow.

Figure 3-3   Concrete channel with trapezoidal cross section.

Let's assume we have a constant rate of flow Q in a channel like that shown in Figure 3-3. If the gradient is varied, then normal depth attained by the flowing water will vary as well. If the gradient is reduced, the velocity is reduced and normal depth is increased and if the gradient is increased (steeper slope) the resulting depth of flow will be shallower. In each case, depth is related to cross sectional area which varies with velocity in accordance with the Continuity Equation, Equation 2-11.

Critical Depth

Now, let's define a quantity called *specific energy*, E, defined as $E = D + v^2/2g$ or total energy head above the channel bed. If we plot values of E against corresponding values of D resulting from variations in depth for the same value of Q we will obtain a curve like that shown in Figure 3-4.

Close inspection of the Specific Energy diagram reveals that for very low velocity with very high depth, the specific energy E approaches the value of D. For high velocity, depth approaches zero meaning that E is composed almost entirely of $v^2/2g$. One particular value of D, called *critical depth*, designated D_c, results in the minimum value of E. Critical depth is a theoretical concept depending only on the channel shape and the flow quantity Q. It does not depend on the roughness of the channel lining nor the slope of the channel. Generally, flow depths greater than critical depth represent more tranquil flow called *subcritical*, while depths below D_c represent more rapid flow called *supercritical*.

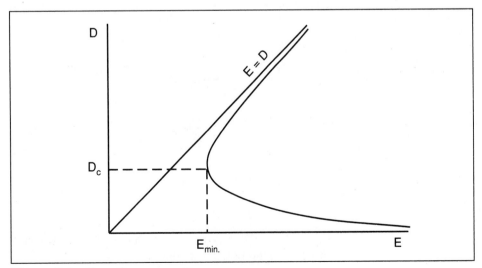

Figure 3-4   Specific energy diagram.

The velocity of the water at critical depth is called *critical velocity*. When a channel has a slope that causes normal depth to coincide with critical depth, the slope is called *critical slope*. Critical depth is computed by use of the following relationship which occurs only at minimum specific energy, E:

$$\frac{a^3}{B} = \frac{Q^2}{g}, \tag{3-1}$$

where a = cross-sectional area of channel, ft² (m²),
     B = top width of channel, ft (m),
     Q = flow rate, cfs (m³/s),
     g = acceleration of gravity, 32.2 ft/s² (9.81 m/s²).

Equation 3-1 applies to all channel shapes and must usually be solved by trial and error. However, Equation 3-1 is valid only if the flow is gradually varied or parallel to the channel bottom and the channel slope is small (less than 8%). For rectangular channels, Equation 3-1 reduces to

$$D_c = \sqrt[3]{\frac{Q^2}{B^2 g}} \tag{3-2}$$

For any given channel and flow rate Q there is a corresponding critical depth. Usually, normal depth in a channel is not equal to critical depth. Only one particular gradient results in a normal depth equal to critical depth.

The most common occurrence of critical depth is in *gradually varied flow* in which the flow depth varies over a certain distance due to a change in one or more of the channel attributes. The change could be in channel gradient or cross section shape. As the depth varies, critical depth can be attained within the transition region. For instance, Figure 3-5 shows a profile of a channel transition from subcritical flow to supercritical flow. Depth of flow is at critical depth near the transition point. Notice in Figure 3-5 that theoretical critical depth does not change when the channel slope changes.

The concept of critical depth, although purely theoretical, has many applications in open channel hydraulics. Two major applications of the principle are flow control and flow measurement. We will refer to critical depth later in our analysis of culverts.

Many special conditions can and do occur in open channel hydraulics which require special mathematical analysis, some of which are depicted in Figure 3-6. However, these analyses are beyond the scope of this text. We will focus on simple channel design considerations in which, under most circumstances, normal depth is assumed in the channel.

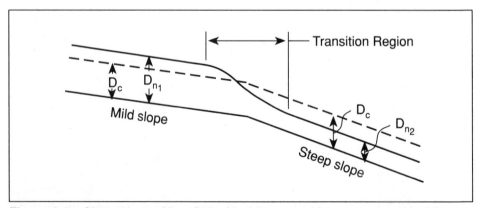

Figure 3-5 Channel transition. Subcritical flow transitions to supercritical flow by passing through critical depth.

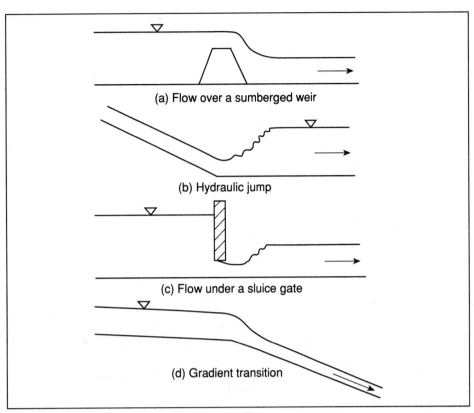

Figure 3-6   Examples of varied flow depth.

3.2   Manning's Equation

As stated previously, several empirical formulas have been devised to calculate normal depth in a uniform channel. The most widely used of these is Manning's Equation, which is stated as follows:

$$Q = a \frac{1.49}{n} R^{2/3} s^{1/2} \qquad \text{(English Units)} \qquad (3\text{-}3)$$

where   Q = flow rate, cfs,
       a = cross-sectional area, ft^2,
       n = roughness factor,
       R = hydraulic radius, ft,
       s = gradient, ft/ft;

$$Q = \frac{a}{n} R^{2/3} s^{1/2} \qquad \text{(SI Units)} \qquad (3\text{-}3a)$$

where   Q = flow rate, m^3/s,
       a = cross-sectional area, m^2,
       n = roughness factor,
       R = hydraulic radius, m,
       s = gradient, m/m.

A quick look at either Equation 3-3 or 3-3a reveals no term for normal depth even though it is used to calculate normal depth. However, this seeming contradiction is answered rather easily: normal depth is calculated indirectly. Manning's Equation is used in an iterative (trial and error) process to find normal depth if Q is known. This process will be shown shortly in Example 3-3.

Figure 3-7 Parameters used in Manning's Equation.

The use of Manning's Equation involves various parameters not yet encountered. First, *roughness factor,* n, is a dimensionless number describing the roughness of the surface of the channel. It is similar to f used in Bernoulli's Equation but not exactly the same; values used for f cannot be substituted for n values. Many design manuals contain charts of Manning n values to be used when working with the Manning Equation. A selection of such n values is listed in Appendix A-1.

Next, *hydraulic radius*, R, is defined as the ratio of cross-sectional area, a, to wetted perimeter, p. Thus,

$$R = \frac{a}{p}.\tag{3-4}$$

Wetted perimeter is the length along the channel cross-section where the flowing water comes in contact with the channel surface as shown in Figure 3-7. Hydraulic radius is not truly a "radius" in a geometrical sense but merely a term defined to give an indication of the hydraulic efficiency of a channel. Thus, a wide shallow channel has a relatively large wetted perimeter compared to cross-sectional area and is therefore less efficient than a square shaped channel with the same cross-sectional area. (See Example 3-1.)

Finally, gradient is a measure of the slope of the profile of the channel as defined earlier. Keep in mind that a slope of 5% is the same as a gradient of 0.05 ft/ft (m/m).

Example 3-1

Problem

Find the hydraulic radius of two channels having the same cross-sectional area but different wetted perimeters as shown. Which channel will carry more flow?

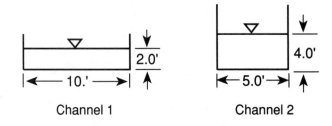

Channel 1   Channel 2

Solution
Channel 1
$$a = (10.) (2.0) = 20. \text{ ft}^2$$
$$p = 2.0 + 10. + 2.0 = 14 \text{ ft}$$
$$R = 20./14 = 1.4 \text{ ft.} \quad \text{(Answer)}$$

Channel 2
$$a = (5.0) (4.0) = 20. \text{ ft}^2$$
$$p = 4.0 + 5.0 + 4.0 = 13 \text{ ft}$$
$$R = 20./13 = 1.5 \text{ ft.} \quad \text{(Answer)}$$

Channel 2 has greater hydraulic radius than does Channel 1. Therefore, it will carry greater flow since according to Manning's Equation Q varies directly with $R^{2/3}$.

Manning's Equation is used to find normal depth in all open channels that can be considered uniform with constant flow. These include concrete channels, grass channels, pipes (flowing part full), swales and streams. In the case of natural streams, however, the assumption of uniform cross section, gradient and roughness is usually not precisely correct. This topic will be discussed in more detail later in this chapter.

3.3   Channel Flow

Charts for the solution of Manning's Equation have been computed for various channel sections. Charts for rectangular and trapezoidal channels are found in Appendix A-3.

Example 3-2

Problem
Find the quantity of flow Q in a concrete rectangular channel having a width of 12.0 feet, a gradient of 0.0200 ft/ft and a normal depth D of 1.58 feet.

Solution
When D is given, Q can be solved directly using Manning's Equation. First, choose the appropriate n value from Appendix A-1. Let n = 0.013 for average concrete with trowel finish. Next, find a:

$$a = BD = (12.0) (1.58) = 18.96 \text{ ft}^2.$$

Next, find p:

$$p = 1.58 + 12.0 + 1.58 = 15.16 \text{ ft.}$$

Next, find R:

$$R = \frac{18.96}{15.16} = 1.25 \text{ ft.}$$

Finally, find Q:

$$Q = a \, \frac{1.49}{n} \, R^{2/3} s^{1/2}$$

$$Q = (18.96) \frac{1.49}{.013} (1.25)^{2/3} (.02)^{1/2}$$

$$Q = 357 \text{ cfs.} \quad \text{(Answer)}$$

The flow velocity can now be easily computed using Equation 2-11:

$$v = \frac{Q}{a} = \frac{357}{18.96} = 18.8 \text{ ft/s.}$$

Example 3-3

Problem

Find normal depth in a 10.0 foot wide concrete rectangular channel having a gradient of 0.0150 ft/ft and carrying a flow of 400. cfs.

Solution

When Q is given, D is found by trial and error (iteration). A value for D is chosen as a guess (or trial) and the corresponding Q value computed. If the computed value of Q is greater than 400. cfs, a second trial using a smaller D is made and if Q is less than 400. cfs, a larger D is chosen. By trial three or four, D can be reasonably approximated.

First, choose the appropriate n value. Let n = 0.013.

Trial 1: Let D = 2.00 feet. Then a = 20.0 ft², p = 14.0 ft and

$$R = \frac{20}{14} = 1.43 \text{ ft}$$

$$Q = (20)\,\frac{1.49}{.013}\,(1.43)^{2/3}\,(.015)^{1/2} = 356 \text{ cfs.}$$

Since 356 is less than 400. cfs, choose a larger value of D.

Trial 2: Let D = 3.00 ft. Then a = 30.0 ft², p = 16.0 ft and

$$R = \frac{30}{16} = 1.88 \text{ ft}$$

$$Q = (30)\,\frac{1.49}{.013}\,(1.88)^{2/3}\,(.015)^{1/2} = 640. \text{ cfs.}$$

The correct value of D lies between 2.00 feet and 3.00 feet. An estimate can be made by interpolating between the two sets of values computed so far. To interpolate, set up a chart of values already determined where x represents the desired D value:

D	Q
2.00	356
x	400.
3.00	640

Interpolation: $\dfrac{x-2}{3-2} = \dfrac{400-356}{640-356}$

$$\dfrac{x-2}{1} = \dfrac{44}{287}$$

$$x - 2 = 0.155$$

$$x = 2.15$$

$$D = 2.15 \text{ ft.}$$

Therefore, an estimate for D is 2.15 ft which must now be checked in trial three.

Trial 3: Let D = 2.15 ft. Then a = 21.5 ft, p = 14.3 ft and

$$R = \dfrac{21.5}{14.3} = 1.50 \text{ ft}$$

$$Q = (21.5)\,\dfrac{1.49}{.013}\,(1.50)^{2/3}\,(.015)^{1/2} = 396 \text{ cfs.}$$

Since 396 cfs differs from 400. cfs by only 1.0%, it is considered a close enough approximation. Therefore, the normal depth has been found to be 2.15 ft. (Answer)

The above calculations can be performed in a convenient tabular format as follows:

First, compute $\dfrac{1.49}{n}s^{1/2}$ because this is the portion of Manning's Equation that will not change from one trial to another.

$$\dfrac{1.49}{n}s^{1/2} = 14.04$$

Next, create a table having each parameter to be computed across the top and then proceed to the trials:

Trial	D (ft)	a (ft²)	p (ft)	R (ft)	R^{2/3} —	Q (cfs)
1	2.00	20	14	1.43	1.27	356
2	3.00	30	16	1.88	1.52	640
3	2.15	21.5	14.3	1.50	1.31	396

The above tabulation could be easily computed using standard spread sheet software such as Lotus 1-2-3 or MS Excel.

Critical Depth

Critical depth in rectangular channels is easily computed using Equation 3-2. Critical depth in a non-rectangular channel is computed from Equation 3-1. Critical depth for trapezoidal channels as well as rectangular channels is also found using the design charts in Appendix A-3. A channel has one particular critical depth for each flow quantity Q. The following example illustrates the use of Equation 3-1.

Example 3-4

Problem

Find critical depth in a trapezoidal channel having a bottom width of 6.00 feet and 2:1 side slopes and a slope of 1.00%. The channel is lined with concrete in good condition and carries a flow of 80.0 cfs.

Solution

The condition of the concrete is irrelevant since D_c does not depend on n. Therefore, first determine the value of Q^2/g which is the right side of Equation 3-1.

$$\frac{Q^2}{g} = \frac{80^2}{32.2} = 199$$

(This value has meaningless units of ft^5 which we will not bother to write in the computations.)

Trial 1: Let D_c = 1.00 feet. Then a = 8.00 ft^2 and a^3 = 512. Geometry of the channel gives B = 10.0 ft.

Therefore, a^3/B = 512/10.0 = 51.2. This value is then compared to the value of 199 already computed for Q^2/g. Since the value of a^3/B is less than 199, choose a higher value for a^3/B.

Trial 2: Let D_c = 2.00 feet. Then a = 20.0 ft^2 and a^3 = 8000. Geometry of the channel gives B = 14.0 feet.

Therefore, $a^3/B = 8000/14 = 571$. Since 571 is greater than 199, the correct value of D_c lies between 1.00 foot and 2.00 feet. An estimate can be made by interpolating between the two sets of numbers computed so far.

D_c	a^3/B
1.00	51.2
x	199
2.00	571

Interpolation: $\dfrac{x - 1}{2 - 1} = \dfrac{199 - 51.2}{571 - 51.2}$

$$x - 1 = \frac{147.8}{519.8}$$

$$x - 1 = 0.284$$

$$x = 1.28$$

$$D_c = 1.28 \text{ ft.}$$

Therefore, an estimate for D_c is 1.28 feet, which must now be checked.

Trial 3: Let $D_c = 1.28$ feet. Then $a = 11.0$ ft^2, $a^3 = 1315$ and $B = 11.12$ ft.

Therefore, $a^3/B = 1315/11.12 = 118$. Since 118 differs from 199 by 41%, it is not considered a close enough approximation. Since 118 is less than 199, choose a higher value of D_c, say $D_c = 1.50$ feet.

Trial 4: Let $D_c = 1.50$ feet. Then $a = 13.5$ ft^2 and $a^3 = 2640$ and $B = 12.0$ feet.

Therefore, $a^3/B = 2460/12 = 205$. Since 205 differs from 199 by 3.0%, it is marginal as a close approximation. Therefore, attempt one more trial with $D_c = 1.49$ feet.

Trial 5: Let $D_c = 1.49$ ft. Then $a = 13.38$ ft^2, $a^3 = 2395$ and $B = 11.96$ feet.

Therefore, $a^3/B = 2395/11.96 = 200$. Since 200 differs from 199 by only 0.5%, it is considered to be a close enough approximation. Therefore, the critical depth has been found to be 1.49 feet. (Answer)

The above calculation can be performed in a convenient tabular format as follows:

First, compute $Q^2/g = 199$. Next, create a table having each parameter to be computed across the top and then proceed to the trials:

Trial	D_c (ft)	a (ft^2)	a^3 —	B (ft)	a^3/B —
1	1.00	8.00	512	10	51.2
2	2.00	20.0	8000	14	571
3	1.28	11.0	1315	11.12	118
4	1.50	13.5	2460	12	205
5	1.49	13.4	2395	11.96	200

As in the case of Example 3-3, this table can be computed using spread sheet software.

3.4   Pipe Flow

When water flows by gravity in a pipe partially full (not under pressure) it conforms to the laws of open channel flow. Normal depth is computed from either Equation 3-3 or 3-3a and critical depth is computed from Equation 3-1. The only difference between pipe flow and rectangular channel flow is the geometry of the cross section.

Principal applications of pipe flow are storm sewers, sanitary sewers and culverts. The following examples illustrate the characteristics of flow in circular pipes.

Example 3-5

Problem

Find the quantity and velocity of flow in an 18-inch diameter concrete pipe flowing one-quarter full. The pipe has a slope of 1.00%.

Solution

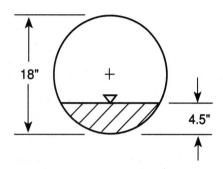

First choose an n value. Say n = 0.015. Next, compute a and p:

To find a, the cross-sectional area of flow is recognized as a segment of a circle, thus from geometric analysis:

a = 0.345 ft².

To find p, compute the arc length of the circle intersected by the water surface. Again, from geometric analysis:

p = 1.57 ft.

Next, compute R:

$$R = \frac{a}{p} = \frac{.345}{1.57} = 0.220 \text{ ft.}$$

Finally, compute Q:

$$Q = a \frac{1.49}{n} R^{2/3} s^{1/2}$$

$$Q = (.345) \frac{1.49}{.015} (.220)^{2/3} (.01)^{1/2}$$

$$Q = 1.25 \text{ cfs.} \quad \text{(Answer)}$$

Also, compute v:

$$v = \frac{Q}{a} = \frac{1.25}{.345} = 3.62 \text{ ft / s.} \quad \text{(Answer)}$$

Example 3-6

Problem

Find the quantity and velocity of flow in an 18-inch diameter concrete pipe flowing half full. The pipe has a slope of 1.00%

Solution

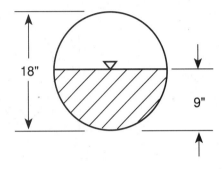

Let n = 0.015. Compute a and p:

$$a = \frac{1}{2}\pi \,(.75)^2$$
$$a = 0.884 \text{ ft}^2$$
$$p = \frac{1}{2}\pi \,(1.5)$$
$$p = 2.36 \text{ ft.}$$

Next, compute R:

$$R = \frac{a}{p} = \frac{.884}{2.36} = 0.375 \text{ ft.}$$

Finally, compute Q:

$$Q = a \,\frac{1.49}{n} \,R^{2/3}s^{1/2}$$

$$Q = (.884) \,\frac{1.49}{.015} \,(.375)^{2/3}(.01)^{1/2}$$

$$Q = 4.56 \text{ cfs.} \quad \text{(Answer)}$$

Also, compute v:

$$v = \frac{Q}{a} = \frac{4.56}{.884} = 5.16 \text{ ft/s.} \quad \text{(Answer)}$$

Example 3-7

Problem

Find the quantity and velocity of flow in an 18-inch diameter concrete pipe flowing 94% full. The pipe has a slope of 1.00%.

Solution

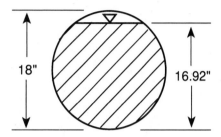

Let n = 0.015. Compute a and p:

The cross-sectional area, a, is found from geometrical analysis to be

a = 1.72 ft^2.

Wetted perimeter, p, also is found from geometrical analysis to be

p = 3.97 ft.

Next, compute R:

$$R = \frac{a}{p} = \frac{1.72}{3.97} = 0.433 \text{ ft.}$$

Finally, compute Q:

$$Q = a\, \frac{1.49}{n}\, R^{2/3} s^{1/2}$$

$$Q = (1.72)\, \frac{1.49}{.015}\, (.433)^{2/3}(.01)^{1/2}$$

Q = 9.78 cfs.   (Answer)

Also, compute v:

$$v = \frac{Q}{a} = \frac{9.78}{1.72} = 5.69 \text{ ft/s.}$$ (Answer)

Example 3-8

Problem

Find the quantity and velocity of flow in an 18-inch diameter concrete pipe flowing just full (but not under pressure). The pipe has a slope of 1.00%.

Solution

In this case, flowing full means the water surface has just reached the top of the pipe and no pressure results. The pipe may still be considered an open channel for this condition.

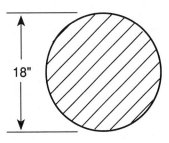

Let n = 0.015. Compute a and p:

$$a = \pi r^2 = \pi(.75)^2 = 1.77 \text{ ft}^2$$

$$p = \pi D = \pi(1.5) = 4.71 \text{ ft.}$$

Next, compute R:

$$R = \frac{a}{p} = \frac{1.77}{4.71} = 0.375 \text{ ft.}$$

Finally, compute Q:

$$Q = a \frac{1.49}{n} R^{2/3} s^{1/2}$$

$$Q = (1.77) \frac{1.49}{.015} (.375)^{2/3} (.01)^{1/2} .$$

$$Q = 9.14 \text{ cfs.} \quad \text{(Answer)}$$

Also, compute v:

$$v = \frac{Q}{a} = \frac{9.14}{1.77} = 5.16 \text{ ft/s.} \quad \text{(Answer)}$$

The four examples above illustrate several facts about pipe flow. First, calculating pipe flow is complex and tedious. For this reason, many design charts (including computer software) have been prepared over the years summarizing the solutions to Manning's Equation for pipe flow. Appendix A-4 presents one such set of charts to be used for problems in this text as well as design projects on the job.

Second, the examples show that the maximum flow in a given circular pipe occurs at a depth of 94 percent of full depth. This odd phenomenon is explained by hydraulic efficiency. As depth approaches the top of the pipe, wetted perimeter (a flow retarder) increases faster than cross-sectional area (a flow increaser). Therefore, the upper 6 percent of a pipe's area contributes nothing to the capacity of the pipe; in fact, it actually decreases the capacity!

Note, however, that when an engineer talks about the capacity of a pipe, he/she generally refers to the Q value when the pipe is flowing full, not flowing at 94 percent depth, despite the fact that flowing full is a little less than the maximum discharge possible in the pipe.

Another characteristic of pipe flow shown in the examples is that velocity is the same for both half and full flow. Velocity generally increases with depth of flow until it reaches a maximum at 94 percent of full flow. Then it decreases until full flow, where it once again reaches the same value as half-full flow.

Figure 3-8 shows a graph of Q versus depth and v versus depth with values from Examples 3-5 thru 3-8 plotted. The pattern that emerges from the plottings illustrates the unique relationships described above for discharge and velocity in circular pipes. A graph depicting these relationships for circular pipes in general is shown in Figure 3-9.

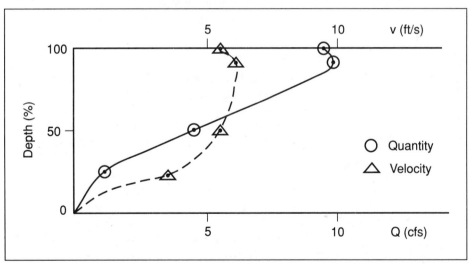

Figure 3-8 Graph of Q and v related to depth of flow for the pipe used in Examples 3-5 through 3-8.

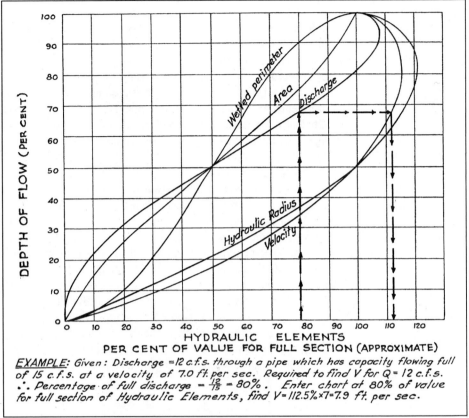

Figure 3-9 Hydraulic elements of circular pipe flow. *(Courtesy of E. Seeley, John Wiley & Sons, Inc. Data Book for Civil Engineers, Vol. 1)*

Critical depth in a pipe flowing as an open channel is computed from Equation 3-1 or taken from the design charts in Appendix A-4. A pipe will have one particular critical depth for each flow quantity Q. The use of Equation 3-1 as an iteration process is quite tedious and, therefore, critical depth is normally taken from design charts.

3.5 Stream Flow

Streams are naturally occurring open channels with varying uniformity of cross-section and roughness. Figure 3-10 depicts two examples of natural streams. Normal depth can be computed for a stream using Manning's Equation when the stream can be considered uniform and when there are no significant obstructions in the stream. Obstructions, such as a bridge crossing, cause a variation in the water surface profile called a *backwater curve*. Figure 3-11 contrasts a backwater profile with a normal depth profile. Backwater curves are plotted by a special computation procedure called a step computation, normally performed by computer. Such computations are beyond the scope of this text.

We will limit our analysis of stream flow to uniform flow in which Manning's Equation is used to compute depth and quantity of flow. Natural streams often flow in a flooded condition; that is, with depth above the top of bank elevation. When flooding occurs, we must still be able to compute the depth even though the roughness factors of the overbanks usually differ from that of the channel of the stream. Figure 3-12 shows a typical stream cross section including overbanks. The usual convention for stream cross sections is that they are drawn looking downstream.

(a) Mountain stream with steep banks.

(b) Lowland stream with flat overbanks.

Figure 3-10   Examples of natural streams.

Figure 3-11 Stream profiles.

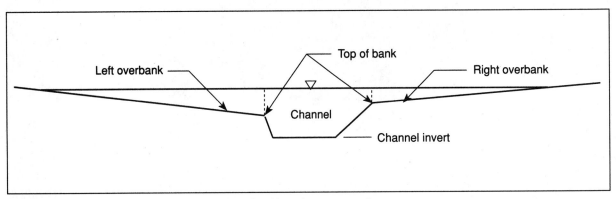

Figure 3-12 Typical steam cross section (looking downstream).

Flow Q is computed separately for all three components of the stream: left overbank, channel and right overbank. Wetted perimeter for the channel is simply the length along the ground in the cross section from top of bank to top of bank. When the water level rises above the top of bank, wetted perimeter does not increase. The following example illustrates the computation of depth in a stream.

Example 3-9

Problem

Find the water surface elevation for the stream depicted below having a flow of 425 cfs and slope of 0.400%. Roughness is described as follows:

      Left Overbank: heavy weeds, scattered brush,
      Channel: fairly regular section, some weeds, light brush on banks,
      Right Overbank: light brush and trees.

Stream Cross Section
Scale: 1" = 10'

Solution

First, determine n values. Using Appendix A-1, n values are estimated as follows:

Left Overbank: n = 0.060,

Channel: n = 0.042,

Right Overbank: n = 0.070.

Next, compute $\dfrac{1.49}{n}s^{1/2}$ for each stream component.

Left Overbank: $\dfrac{1.49}{n}s^{1/2} = \dfrac{1.49}{.06}(.004)^{1/2} = 1.57$

Channel: $\dfrac{1.49}{n}s^{1/2} = \dfrac{1.49}{.042}(.004)^{1/2} = 2.24$

Right Overbank: $\dfrac{1.49}{n}s^{1/2} = \dfrac{1.49}{.07}(.004)^{1/2} = 1.35$

Finally, set up a table having each parameter to be computed listed across the top as shown in Table 3-1, and then proceed to the trials.

Table 3-1
Normal depth computation for stream with overbanks

| | | Left Overbank | | | | | Channel | | | | | Right Overbank | | | | | Total |
Trial	Elev. (ft)	a (ft²)	p (ft)	R (ft)	R²/³ —	Q (cfs)	a (ft²)	p (ft)	R (ft)	R²/³ —	Q (cfs)	a (ft²)	p (ft)	R (ft)	R²/³ —	Q (cfs)	Q (cfs)
1	524	—	—	–	—	—	9	8.3	1.08	1.06	21	—	—	—	—	—	21
2	526	—	—	–	—	—	25.3	12	2.11	1.64	93	3	6	.5	.63	3	96
3	528	10	10.2	.98	.99	15	42.9	12	3.58	2.34	225	26.5	17.5	1.51	1.32	47	287
4	530	40	21	1.9	1.54	97	60.5	12	5.04	2.94	399	73	29.5	2.47	1.83	180	676

The elevation corresponding to Q = 425 cfs can now be estimated by interpolating between trials three and four. Let x = water surface elevation.

Elev.	Q
528	287
x	425
530	676

$$\frac{x-528}{530-528} = \frac{425-287}{676-287}$$

$$\frac{x-528}{2} = \frac{138}{389}$$

$$x - 538 = 0.710$$

$$x = 528.71. \quad \text{(Answer)}$$

Therefore, 528.71 is taken as the water surface elevation corresponding to Q = 425 cfs.

Note: in computing the trials, the areas, a, are cumulative; that is, each value for area includes the area for the preceding trial.

Also, note that all wetted perimeters for the channel for trial two and on are 12 feet. This is because as the water gets above the top of bank, no more contact is made between the water and the ground. For example, in trial three, a and p are shown as follows:

Even though water does not actually flow with vertical sides (unsupported) as shown above, it is assumed to take on such shape purely for computational purposes. Also, note that the information produced by the trials can be used to plot a graph of discharge, Q, versus water surface elevation. Such a graph is called a *stream rating curve* and is shown in Figure 3-13.

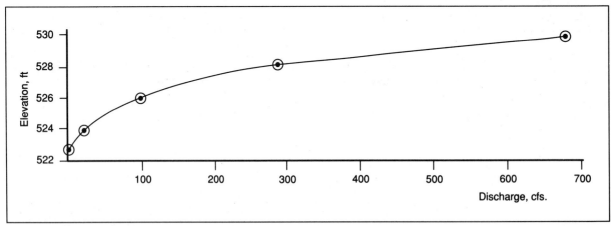

Figure 3-13   Stream rating curve.

3.6 Culvert Flow

Culverts are structures intended to convey a stream through an obstruction such as a road or railroad embankment. The culvert essentially is a tube placed in the embankment allowing the water to flow under the road or railroad with no interference between the two. Figure 3-14 shows a typical culvert/embankment arrangement. Headwalls or wingwalls generally are placed at the upstream and downstream ends to assist grading and hydraulic efficiency. Culverts are made in several cross sections as shown in Figure 3-15 below.

When flow in a stream (or channel) encounters a culvert, it experiences a constriction that causes a change in flow depth. The flow of water in this case can be likened to the flow of traffic from a four-lane highway into a two-lane tunnel. As the traffic enters the tunnel it must squeeze together, thus causing a slow down which affects the cars approaching the tunnel, not the cars traveling through the tunnel. As soon as each car gets into the tunnel, traveling is much easier and it speeds up. When the traffic emerges from the other end, usually traveling is even easier and it speeds up a little more.

Generally speaking, the hydraulics of a culvert behave in a similar way. Water backs up as it is waiting to get into the culvert, and once inside it speeds up. Because of the continuity equation (Equation 2-11), the slow moving water upstream of the culvert has

Figure 3-14 Typical culvert through a road embankment.

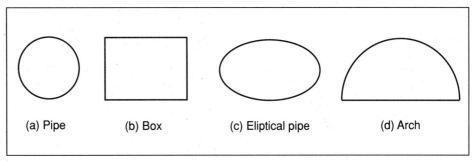

(a) Pipe   (b) Box   (c) Eliptical pipe   (d) Arch

Figure 3-15 Typical culvert cross sections.

an increased depth and the faster water in the culvert is shallower. Figure 3-16 shows a typical pattern of flow through a culvert. It also serves to define many of the specialized terms used in culvert analysis.

Figure 3-17 depicts the principle hydraulic components of culvert flow. Reference will be made to the diagrams in Figure 3-17 later as culvert flow is described in more mathematical terms.

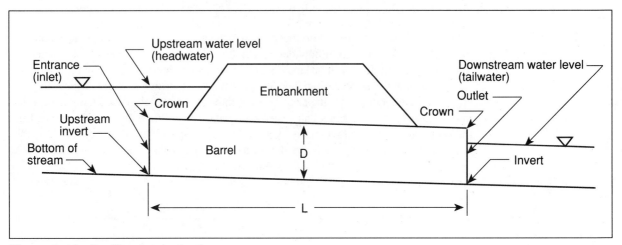

Figure 3-16 Profile of typical culvert.

a) Submerged outlet

b) Unsubmerged outlet

Figure 3-17 Hydraulic components of culvert flow.

Although culvert flow generally behaves as described above, there are several variations of the flow pattern due to specific circumstances. Factors which affect the flow through a culvert are:

1. The size of the opening (cross-sectional area)
2. Entrance geometry
3. Length of the culvert
4. Roughness of the culvert
5. Slope
6. Downstream depth of flow (tailwater)

Downstream depth, also called *tailwater depth*, can be very important in determining the flow pattern. Tailwater depth depends in general on the characteristics of the stream and is usually independent of the culvert. Tailwater depth generally is determined by computing normal depth for the stream as it exists downstream of the culvert.

Several types of culvert flow have been identified by varying the factors listed above. For instance, tailwater depth might be above or below the crown of the culvert. Also, the culvert might be relatively short or long; a long culvert exerts more friction loss on the flow. Furthermore, the culvert slope might be subcritical or supercritical. We can see that many combinations of conditions are possible and each combination results in a different flow pattern through the culvert. However, regardless of the details of the flow pattern, only one part of the pattern is of primary concern in culvert analysis: upstream depth.

It is the upstream water depth, called *headwater depth*, that provides the potential energy to drive the water through the culvert and we will see that this depth becomes a measure of the capacity or adequacy of a given culvert. Downstream (tailwater) depth is determined by the stream while upstream (headwater) depth is determined by the culvert. In this section, we will learn how to compute headwater depth for a variety of flow and culvert factors.

When water flows through a culvert, the headwater depth generally is greater than the tailwater depth. The increase in upstream depth is caused by the constriction inherent in the culvert as we have seen already. However, experimentation has shown that the constriction can take place either at the entrance (upstream end) of the culvert or at the outlet (downstream end) depending upon which of the previously listed factors prevails. If constriction occurs more at the entrance than the outlet, the culvert is said to be operating under *inlet control* whereas if the outlet end creates the greater constriction the culvert is operating under *outlet control*.

Inlet Control

Inlet control occurs when it is harder for the stream flow to get through the entrance of the culvert than it is to flow through the remainder of the culvert and out again into the stream. Critical factors in inlet control are cross-sectional area of the inlet and inlet geometry.

Outlet control occurs when it is harder for the stream flow to negotiate the length of the culvert than it is to get through the entrance in the first place. Critical factors in this case are tailwater depth and the slope, roughness and length of the culvert barrel.

To compute the headwater depth for a culvert operating under inlet control, only the flow rate Q and the entrance size and shape must be taken into consideration. All other factors such as barrel length, roughness, slope and tailwater depth are not important. We will see later how to determine whether or not the given culvert is operating under inlet or outlet control.

Figure 3-18 shows examples of culverts operating under inlet control. In each case, the value HW represents the depth of the headwater measured from the invert (lowest point) of the entrance to the water level. It is HW that we seek to determine in evaluating a culvert operating under inlet control.

Figure 3-18 Examples of culverts operating under inlet control. The inlet can be either unsubmerged or submerged but the outlet is always unsubmerged. *(Courtesy of Federal Highway Administration, Hydraulic Charts for the Selection of Highway Culverts.)*

Note: HW actually represents the vertical distance from the entrance invert to the energy grade line of the headwater, but in most cases the energy grade line is so close to the water surface that they may be taken as being the same.

Design charts are used to determine HW in order to avoid the difficult mathematics involved. The charts, reproduced in Appendix B-1 and B-2, have been developed through research and testing. To use the charts, we must know the stream flow (also called discharge), the size and shape of the culvert entrance cross-section, and the geometry of the entrance end of the culvert barrel.

The entrance cross section generally is either circular, box, oval or arch as shown in Figure 3-15. Entrance geometry relates to the shape of the end of the barrel at the entrance. Figure 3-19 depicts the four usual types of entrance geometry used for circular and oval pipe culverts.

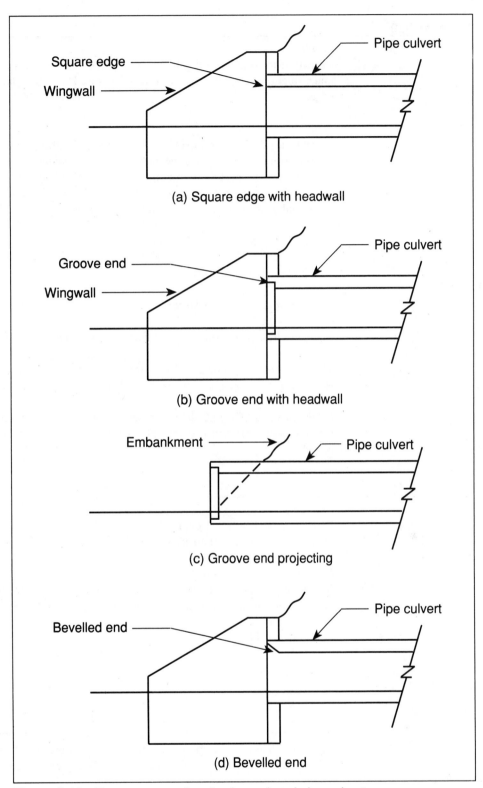

Figure 3-19   Entrance types for circular and oval pipe culverts.

Entrance geometry has a significant effect upon culvert capacity for culverts operating under inlet control despite its minor appearance. The ease with which water can flow past the entrance edge and into the pipe determines the entrance head loss. The smaller the entrance loss the greater the overall capacity.

Example 3-10

Problem

Determine the headwater depth HW of a 36-inch diameter concrete pipe culvert operating under inlet control with a stream flow of 65 cfs. The culvert has a headwall and grooved edge.

Solution

Select Chart 2 (Concrete Pipe Culverts with Inlet Control) in Appendix B-1. Find the pipe diameter on the left-hand scale and find the discharge on the middle scale. Using a straight-edge, connect the two points just determined and extend the line to the lefthand scale of the group of three scales on the right. This is Scale (1). Finally, project the line horizontally from Scale (1) to Scale (2) and read a value of HW/D of 1.57. Then

HW = (1.57) D

HW = (1.57) (3)

HW = 4.71 ft.   (Answer)

Thus, the headwater depth is 4.71 feet which is above the crown of the culvert pipe.

The cross-section geometry of the stream was not needed in this case. Neither was the slope, length or roughness of the pipe. Why?

If the elevation of the headwater is desired, you would just add HW to the elevation of the entrance invert. Thus, if the entrance invert is 450.00, then the headwater elevation is 454.71.

Example 3-11

Problem

Determine the headwater depth of a 3 foot x 6 foot concrete box culvert operating under inlet control with a stream flow of 80 cfs. The culvert has wingwalls at 45° flare.

Solution

Select Chart 1 (Box Culverts with Inlet Control) in Appendix B-1. Find the height of the box (3 feet) on the left-hand scale. Next, compute the value Q/B:

$$\frac{Q}{B} = \frac{80}{6}$$

$$\frac{Q}{B} = 13.3 \text{ cfs/ft.}$$

Find 13.3 cfs/ft on the middle scale and connect the two points just determined and extend the line to Scale (1) of the three scales on the right side of the chart. Since the wingwall flare falls between 30° and 75° use Scale (1) and read the value HW/D of 0.91. Then

HW = (.91) D

HW = (.91) (3)

HW = 2.73 ft.   (Answer)

Thus, the headwater depth is 2.73 feet which is below the crown of the box culvert. If the elevation of the inlet invert is 450.00, then the elevation of the headwater is 452.73.

Outlet Control

Outlet control is a different and more complex procedure. To compute the headwater, depth, barrel size, shape, slope and roughness, as well as tailwater depth, must be known. Figure 3-20 shows examples of culverts operating under outlet control. In each case, the headwater elevation is determined by adding total head H to the hydraulic grade line elevation at the outlet end of the culvert.

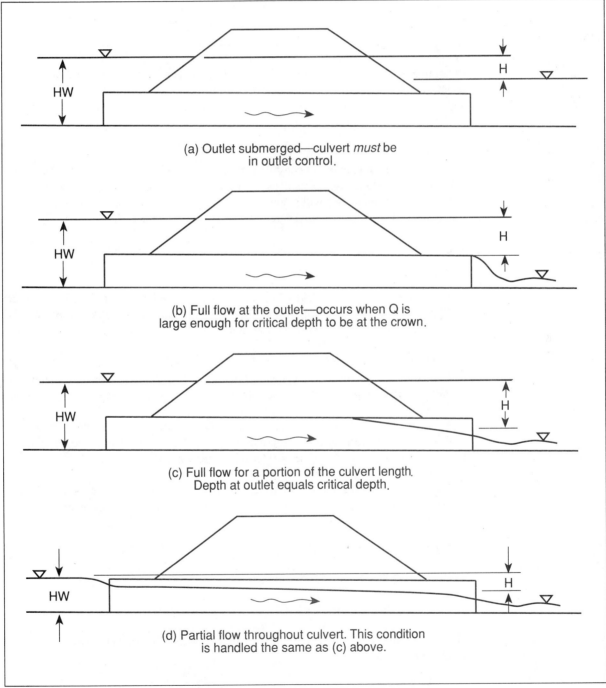

Figure 3-20   Examples of culverts operating under outlet control. *(Courtesy of Federal Highway Administration, Hydraulic Charts for the Selection of Highway Culverts.)*

When tailwater depth is above the crown of the culvert, the hydraulic grade line is at the water level and H is added directly to the tailwater elevation as depicted in Figure 3-17(a). However, when tailwater depth is below the culvert crown, the hydraulic grade line elevation is estimated to be half-way between critical depth D_c and the top of the culvert. Thus, the distance from the invert to the hydraulic grade line (TW') is computed as

$$TW' = (D_c + D)/2. \tag{3-5}$$

So, headwater elevation is equal to the invert elevation plus TW' plus H. It must be noted that when tailwater is below the outlet crown, the depth of flow at the outlet is equal to critical depth as shown in Figure 3-17(b).

Example 3-12

Problem
Determine the headwater depth of a 3 foot x 6 foot concrete box culvert operating under outlet control with a stream flow of 80 cfs. The culvert is 100. feet long and has wingwalls at 45° flare and a square edge at the crown. The slope of both the stream and the culvert is 0.200% and the elevation of the upstream invert is 450.00.

Solution
Because the culvert is in outlet control, analysis begins at the outlet end. First, determine tailwater depth. Normally, this is done by using Manning's Equation as in Example 3-9. In this case, for brevity, we will assume that the depth of flow in the downstream channel was computed to be 3.20 feet. Therefore, the tailwater depth is 3.20 feet.

Since tailwater depth is above the culvert crown, it is not necessary to determine critical depth.

Next, choose Chart 8 in Appendix B-2 to determine the value of H as defined in Figure 3-17(a). However, before using the chart, determine entrance loss, k_e. Appendix B-3 indicates $k_e = 0.4$ for the culvert in question.

Now, locate the cross sectional area of 18 ft² on the Area scale. Next, locate the length of 100 feet on the length scale and connect the two points just determined. Note the point where the connecting line crosses the turning line.

Next, locate the discharge of 80 cfs on the Discharge scale. Draw a line from the discharge point to the turning line point and extend the line to the Head scale and read the head value H which in this case is 0.63 feet. Therefore, H = 0.63 feet.

Finally, calculate headwater elevation by adding (H + TW) to the elevation of the outlet invert. The outlet invert is computed by subtracting Ls from the inlet invert.
Thus, the outlet invert elevation is

$450 - (100) (.002) = 449.80.$

Therefore, the headwater elevation is

$449.80 + 0.63 + 3.20 = 453.63.$

Headwater depth HW is then computed by subtracting the inlet invert elevation from the headwater elevation:

HW = 453.63 − 450.00

HW = 3.63 ft.   (Answer).

Example 3-13

Problem

Determine the headwater depth of a 3 foot x 6 foot concrete box culvert operating under outlet control with a stream flow of 80.0 cfs as in Example 3-12, except that tailwater depth is below the culvert crown (TW = 1.00 feet). All other parameters are identical to those in Example 3-12.

Solution

Since tailwater depth is below the culvert crown, critical depth must be determined.

> *Note:* To find critical depth in a box culvert, use the channel charts in Appendix A-3 choosing a rectangular channel with the same width as the culvert.

Using Chart 5 in Appendix A-3, critical depth is found to be $D_c = 1.50$ feet. The position of the hydraulic grade line (TW') is then computed as

$$TW' = \frac{(D_c + D)}{2}$$

$$TW' = \frac{(1.5 + 3)}{2}$$

$$TW' = 2.25 \text{ ft.}$$

Since TW' is greater than TW, TW' will be used to compute headwater elevation. Total head H is determined to be 0.63 ft using Chart 8 of Appendix B-2 as in Example 3-12.

Now, compute headwater elevation by adding (H + TW') to the elevation of the outlet invert. Therefore, headwater elevation is

449.80 + H + TW'

449.80 + 0.63 + 2.25 = 452.68.

Headwater depth is then computed by subtracting the inlet invert elevation from the headwater elevation:

HW = 452.68 − 450.00
HW = 2.68 ft.   (Answer)

In analyzing a given culvert, we usually do not know whether it operates in inlet or outlet control. Various mathematical methods can be used to determine the type of control based upon the specific parameters involved. However, an easier method is available: Simply calculate upstream depth HW assuming inlet control and then again compute HW assuming outlet control and then compare the two answers. Whichever value of HW is greater indicates the prevailing control and becomes the answer to the analysis.

Therefore, the proper method used to determine headwater depth for any culvert problem is to analyze separately both inlet and outlet control and then choose as the answer the greater value of HW from the two values determined.

The above methodology using culvert charts is based upon the assumption that tailwater depth is normal depth in the downstream channel, implying that no significant obstructions are located downstream of the culvert. If obstructions are located downstream, then stream depth at the outlet of the culvert will be greater than normal depth, and the value of TW will be greater, thus affecting the determination of HW.

In such cases, a different method of analysis must be used. Tailwater depth must be computed using a backwater computation as defined in Section 3.5. The usual method used when backwater computations are needed is the computer program HEC-2 developed by the U.S. Army Corps of Engineers. The use of backwater computations or of HEC-2, however, are beyond the scope of this text. Refer to Appendix E for a list of applicable computer software.

PROBLEMS

1. A 6.0-foot wide rectangular channel carries a flow of 56 cfs at a depth of 1.50 feet. What is the average velocity of the flow?

2. A channel conveys a discharge of 210 cfs at a velocity of 5.45 ft/s. Find the cross-sectional area of the channel.

3. A trapezoidal channel with bottom width of 4.00 feet and side slopes of 2 horizontal : 1 vertical carries a flow depth of 2.25 feet at an average velocity of 7.32 ft/s. What is the flow, Q, in the channel?

4. Find normal depth for a discharge of 125 cfs in a trapezoidal channel with bottom width of 3.00 feet and side slopes of 2 horizontal : 1 vertical. The channel is lined with riprap (n = 0.024) and has a slope of 1.00%. Use Manning's Equation.

5. Find normal depth for a discharge of 350. cfs in a rectangular channel with bottom width of 20.0 feet. The channel is concrete lined (n = 0.013) with a slope of 0.75%. Use Manning's Equation.

6. A 36-inch diameter pipe flowing half full conveys a discharge of 15 cfs. What is the velocity of the flow?

7. Find the critical depth for the channel in Problem 4.

8. Water is flowing in a 5.0-foot wide rectangular channel with a quantity of 100. cfs. Plot a graph of D_n vs E by choosing a variety of flow depths D_n and for each flow depth computing the corresponding value of E using Equation 2-11. Use the graph you plotted to determine critical depth.

9. Find the capacity of a 15-inch RCP with n = 0.015 and s = 1.25%.

10. Find the capacity of a 36-inch RCP with n = 0.012 and s = 0.80%.

11. For the stream cross section shown below, find the water surface elevation for a flow of 650 cfs. The stream has a gradient of 0.110% and the following characteristics:

    Channel: Fairly regular section, some weeds, light brush on banks.

    Overbanks: Heavy weeds, scattered brush.

Inv. 226.30

Datum

Stream Cross Section
Scale: 1" = 10'

12. Find the headwater elevation for a stream flow of 65 cfs for the culvert shown below. Tailwater depth is 2.00 feet. The culvert has wingwalls with 45° flare and a square edge.

13. Find the headwater elevation for a stream flow of 100. cfs for the culvert shown below. Tailwater depth is 4.00 feet. The culvert has wingwalls with 45° flare and a square edge.

14. Find the headwater elevation for a stream flow of 140 cfs for the culvert shown below. Tailwater depth is 2.50 feet. The culvert has no wingwalls and has a square edge.

15. The culvert shown below conveys 150. cfs flowing in a stream with cross section as shown. Find the headwater elevation. The culvert has wingwalls with 45° flare and a square edge.

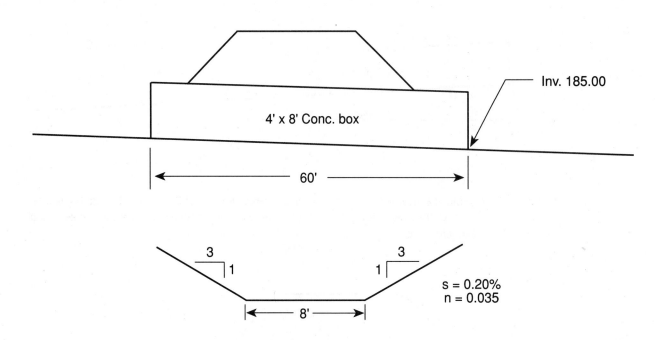

FURTHER READING

Brater, E. F. and King, Horace. (1976). *Handbook of Hydraulics*. New York: McGraw-Hill.

Chow, V. T. (1985). *Open Channel Hydraulics*. New York: McGraw-Hill.

Herr, L. (1965 revision). *Hydraulic Charts for the Selection of Highway Culverts, Hydraulic Engineering Circular No. 5*. Washington, DC: U.S. Dept. of Commerce, Bureau of Public Roads.

Linsley, R. K. and Franzini, J. B. (1992). *Water Resources Engineering* (4th ed.). New York: McGraw-Hill.

United States Department of the Interior, Bureau of Reclamation (1977). *Design of Small Dams* (2nd ed. revised reprint). Washington U.S. Government Printing Office.

4

RUNOFF

Before rainwater can flow in a stream or channel or culvert as described in Chapter 3, it must first descend from the sky and make its way across the surface of the land and accumulate into a concentrated form.

In this chapter, we will see how to calculate the quantity of flow, Q, that results from the rainfall and which must be conveyed by the hydraulic structure to be designed. The value of Q resulting from rainfall is called *runoff* and depends on several factors including the amount of rainfall, the size of the area upon which the rainfall lands and the nature of the ground over which the rainwater flows. Each of these factors will be discussed in detail in the sections that follow.

OBJECTIVES

After completing this chapter, the reader should be able to:
- Describe the hydrologic cycle
- Delineate a drainage basin on a topographic map
- Calculate peak runoff by the Rational Method
- Calculate peak runoff by the SCS Method
- Calculate a runoff hydrograph by the SCS Method

4.1 Hydrologic Cycle

Nature is a great recycler and water is a prime example. The water that flows in streams comes from the ocean and is returned again in a constant cycle called the *hydrologic cycle*.

Although complex in its functioning, the hydrologic cycle can be simply explained as the following steps:

1. Water evaporates from the oceans and lakes of the earth.
2. The evaporated water vapor forms into clouds.
3. The clouds move through the atmosphere in global weather patterns.
4. The water vapor condenses and precipitates in the form of rain or snow or hail.
5. The rain lands on the ground and flows overland to the streams.
6. The streams flow to rivers and eventually into the oceans and lakes.

Figure 4-1   Schematic diagram of the hydrologic cycle.

A schematic diagram of the hydrologic cycle is shown in Figure 4-1.

One aspect of stream flow is not explained by the above oversimplified description of the hydrologic cycle and that is the constancy of the flow we observe in rivers and large streams. You might think that streams would flow only when it rains, but experience contradicts this notion.

In fact, streams and rivers are fed by rainfall in three ways. Figure 4-2 shows a typical stream cross section and illustrates the fate of rainwater when it lands on ground near the stream. Some of the rain is lost immediately to evaporation and *evapotranspiration*, the loss of water vapor from plants to the atmosphere; some flows by gravity over the surface of the ground and eventually into the stream; and the remainder infiltrates into the ground. Of the infiltrated water, some flows in underground or subsurface flow by gravity to the stream, and some percolates to the stored underground water body called

Figure 4-2   Typical stream cross-section showing three ways that stormwater reaches the stream.

groundwater. Although the overland flow is relatively quick, reaching the stream in minutes or, at most, hours, the subsurface flow and groundwater flow is much slower, lasting many days. Therefore, after an initial quick surge of overland flow when it rains, streams receive a constant feed of subsurface water throughout the days between rainfalls.

The constant low-level flow in streams due to the subsurface feed is called *base flow* and the quick surge due to overland flow is called *flood flow*. It is flood flow that we are concerned with in the design of hydraulic structures because flood flow represents the greatest volume of water which the structure must handle.

Following any normal rainfall event, flood flow peaks and begins to diminish before subsurface and groundwater flows have a chance to make a significant contribution to the stream. Therefore, the flow rate we call runoff will be assumed to consist entirely of flood flow.

4.2   Drainage Area

In calculating the rate of runoff in a stream resulting from a rainfall event, we must first determine the size of the area over which the rain falls. For every stream, there is a well defined area of land which intercepts the rainfall and transports it to the stream. The area of land is called the *catchment area, watershed* or *drainage basin*. These three terms generally are used interchangeably. Figure 4-3 shows a typical drainage basin for a stream. All rainwater which lands within the drainage basin makes its way to the stream while all rain landing outside the drainage basin makes its way away from the stream and into some other stream.

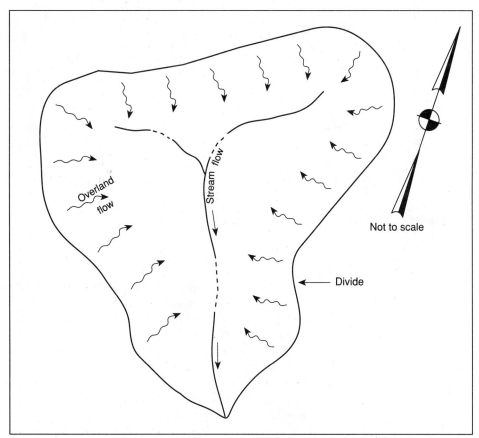

Figure 4-3   Plan view of typical drainage basin. Overland flow is directed by gravity away from the divide and toward the steam.

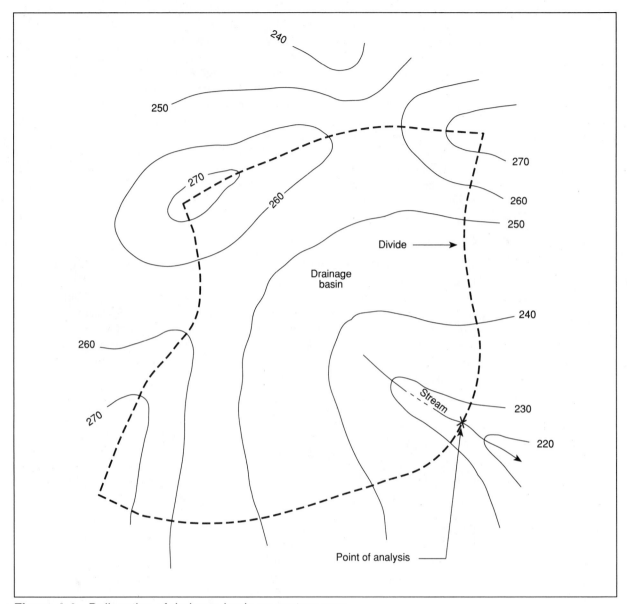

Figure 4-4 Delineation of drainage basin on contour map.

The imaginary line which outlines the boundary of the drainage basin is called the *basin divide* and is determined by the topography of the land. Delineating the basin divide is done on a contour map of the land surrounding the stream and is the first step in computing runoff.

Figure 4-4 shows a simplified contour map with a drainage area delineated upon it. The first step in delineating the drainage basin is deciding the point on the stream where the basin starts. The starting point is called the *point of analysis* and can be anywhere along the stream. Normally, the point of analysis is chosen at the location of a proposed hydraulic structure, such as a culvert.

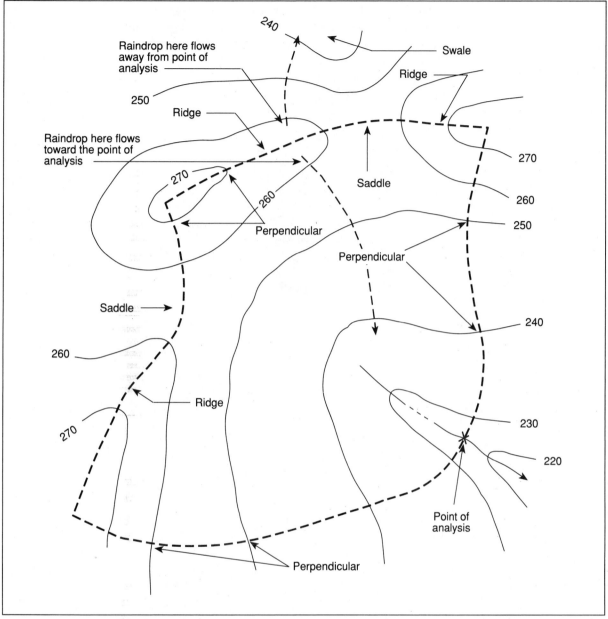

Figure 4-5   Principles in delineating a drainage basin.

The basin divide in Figure 4-5 illustrates the major principles in delineating a drainage basin. These principles are summarized below.

In delineating a drainage basin:

1. Draw the divide perpendicular to contour lines (when the contour lines represent a slope.)
2. Draw the divide along a ridge and across a saddle.
3. Never draw the divide along or across a swale.
4. Draw the divide between and parallel to two contour lines of the same elevation.
5. When in doubt about your line, test it by imagining a drop of rain landing near the line; then trace the runoff path taken by the drop. If the drop flows toward the point of analysis, it landed inside the basin. (When water runs downhill, it travels perpendicular to the contour lines.)

The principle land features involved in basin delineation are shown in Figure 4-6. A *swale* is characterized by bent contour lines that point uphill while a *ridge* is the opposite; the contour lines point downhill. A *saddle* is the transition between two ridges and two swales.

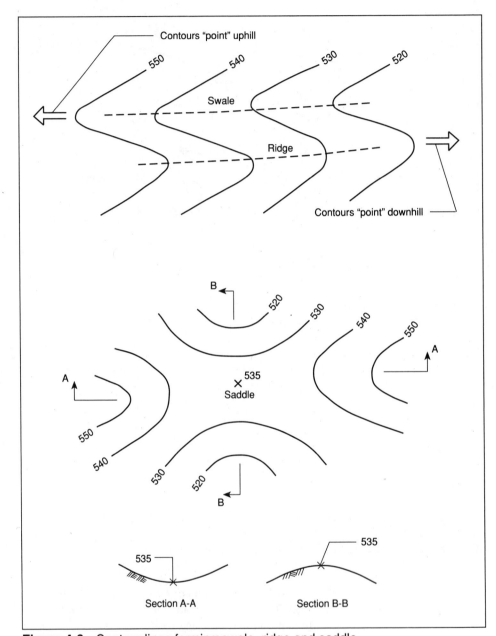

Figure 4-6 Contour lines forming swale, ridge and saddle.

Figure 4-7   Using a planimeter to measure a drainage area.

After you have delineated the drainage basin, your next step is to measure the area of the basin. The units normally are acres in the English system and square meters in the SI system. The area measurement is of the horizontal plane contained within the delineation; not the actual surface of the ground.

Measure the area using one of two methods: (1) *planimeter* or (2) approximate geometric shapes. Figure 4-7 shows a planimeter in use. When using a planimeter, always measure the area three times and then compute an average. Figure 4-8 shows a typical drainage area computed by both planimeter and geometric shapes. If the area is within 5% of the actual area, it is accurate enough for design. This is because delineating a drainage area is from the outset an approximation; the actual area can never be known with absolute precision.

Measure by triangles:
1.   3,360,000 s.f.
2.   9,950,000 s.f.
3. 10,198,750 s.f.
4. 19,775,000 s.f.
5.   8,125,000 s.f.
6. 10,115,000 s.f.
TOTAL 61,523,750 s.f.

$$\frac{61{,}523{,}750}{43{,}560} = 1412 \text{ acres}$$

Measure by planimeter:
1. 1423 acres ⎫
2. 1419 acres ⎬ 1423 acres (average)
3. 1426 acres ⎭

Point of analysis

SCALE: 1" = 2000'

Figure 4-8 Computation of typical drainage area.

4.3 Time of Concentration

When the size of the drainage basin has been determined, the next step in finding Q is to compute the *time of concentration*, t_c. Time of concentration is defined as the amount of time needed for runoff to flow from the most remote point in the drainage basin to the point of analysis. The path or route taken by the most remote drop is called the *hydraulic path* and is illustrated in Figure 4-9.

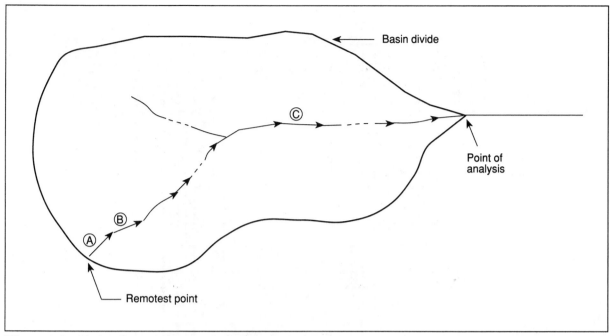

Figure 4-9 Hydraulic path for a typical drainage basin. Initially, runoff travels in overland flow, segment A; then in shallow concentrated flow, segment B; and finally in stream flow, segment C.

The hydraulic path may not cover the greatest distance, but it takes the greatest time compared to the routes of all other drops in the drainage basin. The time is determined by adding all the individual flow times for the different types of flow as the drop makes its way toward the point of analysis. Therefore,

$$t_c = t_1 + t_2 + t_3 + \dots + t_n$$

where t_1, etc. represent the travel times for overland flow, shallow concentrated flow, stream flow and any other type of flow encountered.

Overland flow is usually the first type of flow as the drop starts from the remotest point. It is characterized by sheet flow down a relatively featureless slope similar to the manner in which water flows across pavement. This is the slowest of all types of flow and is computed either by a nomograph (see Example 4-1) or by empirical formula. Typically, overland flow cannot travel more than 300 feet before consolidating into a more concentrated flow.

Shallow concentrated flow occurs when the natural indentations of terrain cause the runoff to form into small rivulets. Since the rivulets are more concentrated, the flow efficiency is increased and therefore the velocity is also increased. Time for shallow concentrated flow is determined by empirical nomograph such as that shown in Figure 4-10. (See Example 4-1.)

Stream flow is usually the last (and the fastest) flow to occur along the hydraulic path. Time for stream flow can be computed using Manning's Equation. (See Example 4-1)

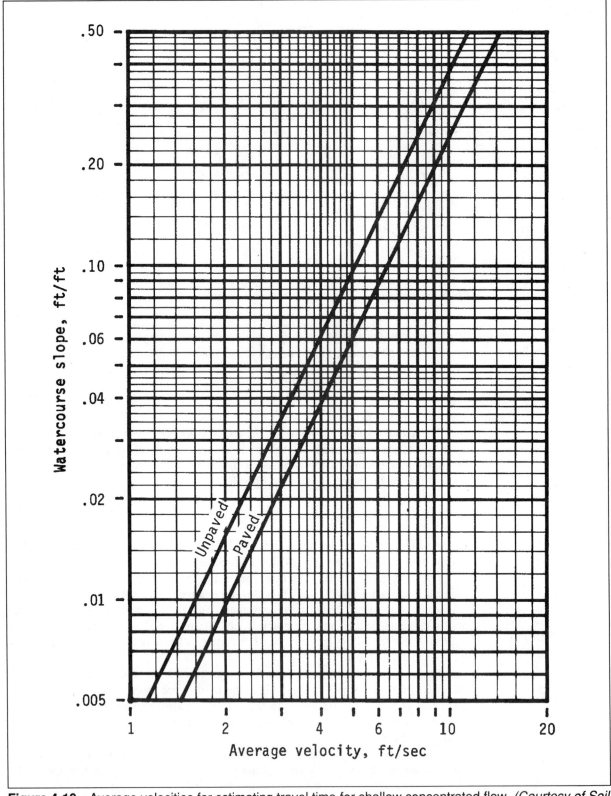

Figure 4-10 Average velocities for estimating travel time for shallow concentrated flow. *(Courtesy of Soil Conservation Service, Technical Release 55.)*

Example 4-1

Problem

Determine the time of concentration for the drainage basin shown in Figure 4-9 having the following conditions:

A. Overland flow: 300' @ 2.5%, average cover
B. Shallow conc. flow: 400' @ 4.0%
C. Stream flow: 4,700' @ 0.3%, average cross section as shown:

Solution

A. Overland flow—using the nomograph in Appendix C-2, overland flow time is found to be 19 minutes. Therefore, t_1 = 19 minutes.
B. Shallow concentrated flow—using the graph in Figure 4-10 for unpaved surface, Velocity = 3.2 ft/s.

$$t_2 = \frac{d}{v} = \frac{400}{3.2}$$

$$t_2 = 125 \text{ s} = 2.1 \text{ min.}$$

C. Stream flow—Since our ultimate goal in finding t_c is to find Q for flood conditions, assume that the stream is filled up to its top of bank. (This assumption is made valid by the fact that as water level rises onto the overbanks, channel velocity does not appreciably increase.)

$$a = 34.5 \text{ ft}^2$$
$$p = 17.2 \text{ ft}$$
$$R = 2.01 \text{ ft}$$

$$v = \frac{1.49}{n} R^{2/3}s^{1/2}$$

$$v = \frac{1.49}{.040} (2.01)^{2/3}(.003)^{1/2}$$

$$v = 3.25 \text{ ft/s}$$

Therefore,

$$t_3 = \frac{d}{v} = \frac{4.700}{3.25}$$

$$t_3 = 1.446 \text{ s} = 24.1 \text{ min.}$$

Finally, t_c is found as the sum of the three individual flow times.

$$t_c = 19 + 2.1 + 24.1$$
$$t_c = 45 \text{ min.} \quad \text{(Answer)}$$

It is important to remember that not all drainage basins will have the same three types of flow along the hydraulic path. For instance, we will see later that some basins have no stream, but may have pavement, pipes or drainage ditches.

Also, in Example 4-1, the implicit assumption was made that all three grades used were uniform. Although such an assumption usually can be made, conditions in the field rarely are so simple. If, for instance, the stream grade was steep for a portion of its length and flatter for the remainder, two separate times might have to be computed to determine total flow time in the stream. This concept also applies to all other flow types.

4.4   Rainfall

Rainfall occurs in haphazard patterns making it very difficult to quantify for design purposes. However, over the past hundred years, reams of data have been compiled on rainfall in the United States. A statistical analysis of this information leads to the determination of an average or typical rainfall.

The result of the statistical analysis is a series of synthetic or theoretical storms categorized by total inches of rainfall and the time it takes for the rain to fall. These categories of storms have been determined for each area of the country. The U.S. Weather Bureau published a report called the *Rainfall Frequency Atlas of the United States* (Technical Paper 40) which shows, through a series of maps, the expected size of storms throughout the country. Later, it was realized that the study did not show enough detail in the mountainous regions west of 103°W longitude. So a follow-up study was prepared by the National Oceanic and Atmospheric Administration (NOAA) to depict more detailed precipitation data for the 11 western states of continental U.S.A. The study, called *NOAA Atlas 2*, is bound in eleven volumes, one for each state, each containing a series of maps depicting the precipitation data. A selection of these maps for both eastern and western states is reproduced in Appendix D-3 for use in determining rainfall at various locations in the continental United States.

The size of a storm is described by the number of inches of rainfall and the duration of the rainfall. Thus, a rainfall event of 5.0 inches over 12 hours is one category while a rainfall of 5.0 inches over 24 hours is another. Even though both events produced the same rainfall, one is more intense and intensity of rainfall is very important in computing runoff.

Probability of occurrence is described by the term *return period* which is the average number of years between two rainfall events which equal or exceed a given number of inches over a given duration. So, for example, a rainfall of 5.0 inches in 24 hours in western Texas has a return period of 10 years whereas a rainfall of 5.0 inches in 12 hours has a return period of 25 years. This is because the latter storm is much more intense and therefore a rarer occurrence. These two storms as depicted in the Rainfall Frequency Atlas are shown in Figures 4-11 and 4-12, respectively.

10-YEAR 24-HOUR RAINFALL (INCHES)

5" RAINFALL

Figure 4-11 Map showing rainfall depths for a 10-year, 24-hour rainfall event. *(Courtesy of U.S. Weather Bureau, Rainfall Atlas of the United States.)*

25-YEAR 12-HOUR RAINFALL (INCHES)

5" RAINFALL

Figure 4-12 Map showing rainfall depths for a 25-year, 24-hour rainfall event. *(Courtesy of U.S. Weather Bureau, Rainfall Atlas of the United States.)*

If you look closely at Figure 4-11, you will see the line representing 5 inches of rainfall passing through western Texas. Then, if you scrutinize Figure 4-12, you will see the 5-inch line passing through western Texas in almost the same pattern as in Figure 4-11. Thus, a 10-year, 24-hour storm delivers the same total rainfall as does a 25-year, 12-hour storm.

After compiling rainfall data for a number of years meteorologists have also been able to establish average patterns of rainfall within a typical storm. Figure 4-13(a) shows a typical pattern of intensity throughout a 24-hour storm. Our common experience reveals that rainfall intensity fluctuates throughout each storm and each storm is different. However, a long-term average produces a single pattern as illustrated in Figure 4-13(b).

a) Typical intensity pattern for one individual storm

b) Intensity pattern based upon long term average

Figure 4-13 Rainfall intensity pattern for 5.0-inch, 24-hour storm in New Jersey.

You might imagine that for a given location in the U.S., a series of such patterns could be drawn, one for each rainfall amount, that is, 1-inch, 24-hour; 2-inch, 24-hour and so on. For each graph thus drawn, the area under the curve would represent the total inches of rainfall for that storm.

Analysis of intensity and duration for an average storm pattern reveals that the storm has one very intense period near the half-way point and less intense periods before and after. If you study Figure 4-13(b) you will see that you could arbitrarily select a period (or duration) of 3 hours in the most intense portion of the storm during which 2.7 inches fall. The intensity during this period is 0.90 inch/hour. Similarly, if you look at the most intense 2 hours of the storm, you would find that 2.5 inches fall giving an intensity of 1.25 inches/hour. Finally, if you look at the most intense 1 hour of the storm, you would find that 2.0 inches fall giving an intensity of 2.0 inches/hour.

These examples suggest a principle: that for smaller durations of time in the most intense portion of a storm the rainfall intensity increases.

The relationship between rainfall intensity and duration for various return periods for a given location in the U.S. is shown on *intensity-duration-frequency (I-D-F) curves* developed by various governmental agencies based upon data from Weather Bureau records. Selected examples of I-D-F curves for various locations in the United States are reproduced in Appendix C-3 for use in solving runoff problems by the Rational Method.

The intensity-duration relationship is central to the Rational Method for determining peak runoff, which is presented in the next section. Therefore, it is important that you study it and make it clear in your mind.

Example 4-2

Problem
In a rainfall event occurring in New Jersey, 4.0 inches of rain falls in a duration of 20 minutes. What is the return period of this storm?

Solution
Using the I-D-F curves for New Jersey in Appendix C-3, enter the graph at a duration of 20 minutes and project a line upward. Then enter the graph at 4.0 inches and project a line across to the right and locate the intersection of the two lines.

Note that the intersection falls on the curve representing the 10-year frequency storm. Therefore, this rainfall event has an approximate return period of 10 years and is referred to as a 10-year storm. (Answer)

4.5   Rational Method

Many methods to compute runoff have been developed over the years, and the first and most enduring of these is the Rational Method. Most methods are based upon empirical relationships among drainage area, time of concentration, rainfall and other factors. However, the Rational Method, introduced in England in 1889, has its genesis in pure reasoning, from which it received its name.

The Rational Method is used to compute the peak runoff, Q_p, following a rainfall event. It makes no attempt to estimate runoff before or after the peak, but simply the one quantity of flow which is greatest.

Originally, the Rational Method formula for peak runoff was given as

$$Q_p = Ai \qquad\qquad (4\text{-}1)$$

where Q_p = peak runoff, cfs
   A = drainage area, acres
   i   = rainfall intensity, in./h

This was based upon a completely impervious drainage basin in which all rainfall is converted to runoff. Later, a proportionality factor, c, called the *runoff coefficient* was added in an attempt to account for infiltration into the ground and for evapotranspiration. So, the formula became

$$Q_p = Aci \tag{4-2}$$

where c is the dimensionless runoff coefficient.

Equations 4-1 and 4-2 are dimensionally balanced in the following way:

$$Ai = acres \times \frac{inches}{hour}$$

$$= 43560 \ ft^2 \times \frac{\frac{ft}{12}}{3600 \ s}$$

$$= \frac{43560 \ ft^3}{43200 \ s}$$

$$= 1.0083 \ ft^3/s$$

Since 1.0083 is nearly equal to 1, it is dropped from the formula.

Equations 4-1 and 4-2 are also consistent with SI units. So, for example, if A is given in m^2 and i in m/s the resulting Q_p will have units m^3/s.

The original reasoning expressed in the Rational Method is described as follows:

1. Consider a drainage basin with area A which is rained on at a constant intensity i for a duration equal to t_c. Assume that no rain falls before or after the downpour of intensity i and duration t_c. Figure 4-14 shows a graphic representation of the rainfall.

Note: To help understand the idea above, remember that in the intensity pattern shown in Figure 4-13, a duration of any length can be considered and then idealized into a constant intensity over that duration. In this case, we are picking a duration which happens to be the same time as t_c.

2. Now picture yourself at the point of analysis watching the runoff go flowing by. As soon as the rainfall starts, some runoff flows by. This early runoff comes from rainfall that landed near the point of analysis. As time goes by runoff increases as rain drops from farther away reach the point of analysis. (But even as these farther raindrops arrive, the close raindrops are still arriving because it is still raining.)

3. Finally, when a time equal to t_c has elapsed, the remotest raindrop arrives at the point of analysis and at this moment the rainfall stops. This then marks the maximum runoff rate and the rate begins declining immediately. After another period equal to t_c, all the water in the drainage basin has run off.

4. To quantify the peak runoff, Q_p, examine Figure 4-14. The total volume of runoff equals the area under the graph of runoff versus time. Thus volume = $\frac{1}{2}(2t_c)(Q_p)$. But remember that total runoff equals total rainfall. Total rainfall is computed simply as the inches of rainfall times the area over which the rain occurred. Thus rainfall volume = (i) (t_c) (A) where (i) (t_c) represents the inches of rainfall.

Therefore, $\frac{1}{2}(2t_c)(Q_p)$ = (i) (t_c) (A) or $Q_p = iA$ which represents the Rational Method for a totally impervious drainage basin.

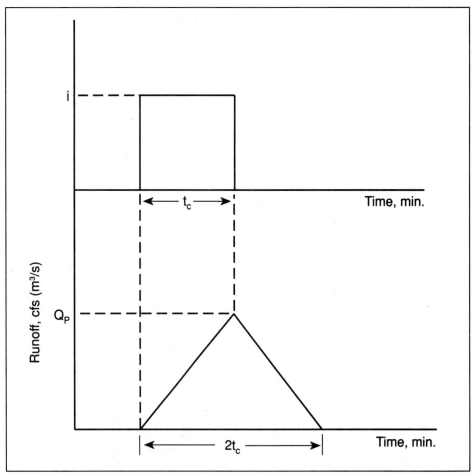

Figure 4-14   Relationship of rainfall to runoff for an impervious drainage basin according to the Rational Method.

5. To account for infiltration and evapotranspiration of runoff, a proportionality constant c was added to the equation. Values of c are empirically determined and correspond to various surface conditions within the drainage basin. Table 4-1 lists some common c values in use today.

6. The fact that rain is falling on the catchment area before and after the duration t_c does not change the derived value of Q_p. The only effect is to change the shape of the graph of runoff versus time.

7. Keep in mind the actual rainfall covers a period of several hours. We have chosen to focus on the most intense portion of the rainfall having a duration of t_c. Any other duration would result in a smaller value of Q_p. If we chose a duration less than t_c the intensity would be greater but we would not include runoff from the entire drainage basin. If we chose a duration greater than t_c the intensity would be less thus reducing mathematically the amount of rainfall.

> **Table 4-1**
> **Typical Runoff coefficients for use with the Rational Method.**

Surface Type	Runoff Coefficient	
	Range of Values	Typical Design
Impervious (Pavement, roofs)	0.75 – 0.95	0.90
Lawns	0.05 – 0.35	0.30
Unimproved (Woods, brush)	0.10 – 0.30	0.20

Note: These values of c are typical for lower intensity storms (up to 25-year return period). Higher values are appropriate for larger design storms.

Note: A more extensive list of c values is presented in Appendix C-1.

Example 4-3

Problem

Compute the peak runoff, Q_p, for a 25-year storm using the Rational Method for a drainage basin located in Pennsylvania (Region 1) and having the following parameters:

1. Area: A = 24 acres

2. Time of concentration:

    a) Overland: average grass surface
        Length = 300 ft
        Slope = 2.0%

    b) Shallow concentrated flow:
        Length = 550 ft
        Slope = 4.0%

    c) Stream:
        Length = 1000 ft
        Slope = 0.4%
        Average cross section:

3. Runoff coefficient:
   a) Impervious, 0.5 acre @ c = 0.90
   b) Grass, 11.5 acres @ c = 0.35
   c) Wooded, 12 acres @ c = 0.25

Solution

To compute Q_p, the values of A, i and c must be determined. In this case, A is known to be 24 acres. To find rainfall intensity, i, we first find time of concentration, t_c.

Time for overland flow, t_1, is found from the nomograph in Appendix C-2:

$t_1 = 19.5$ min.

Time for shallow concentrated flow, t_2, is found from Figure 4-10 using the line marked "Unpaved":

v = 3.2 ft/s

$$t_2 = \frac{d}{v} = \frac{550}{3.2} = 172 \text{ s}$$

$t_2 = 2.9$ min.

Time for stream flow, t_3, is found using Manning's Equation. Assume that the stream is flowing bank full. First, find cross-sectional area, a, and wetted perimeter, p, using the average cross section given:

$a = 27 \text{ ft}^2$
$p = 14.5$ ft.

Next, find hydraulic radius, R:

$$R = \frac{a}{p}$$

$$R = \frac{27}{14.5} = 1.86 \text{ ft.}$$

Next, find v from Manning's Equation:

$$v = \frac{1.49}{n} R^{2/3} s^{1/2}$$

$$v = \frac{1.49}{.032} (1.86)^{2/3} (.004)^{1/2}$$

v = 4.45 ft/s.

Finally, find t_3:

$$t_3 = \frac{d}{v} = \frac{1000}{4.45} = 225 \text{ seconds}$$

$t_3 = 3.7$ min.

Therefore, time of concentration is

$$t_c = t_1 + t_2 + t_3$$

$$t_c = 19.5 + 2.9 + 3.7$$

$$t_c = 26.1 \text{ min.}$$

The time of concentration, 26.1 min, is taken as the duration of rainfall in determining rainfall intensity, i:

Duration = 26.1 min.

Rainfall intensity is found using the I-D-F curves for Pennsylvania (Region 1) shown in Appendix C-3. Use the 25-year curve.

i = 2.65 in/h

Now, find the composite runoff coefficient using the c-values given:

$$c = \frac{(.5)\,(.9) + (11.5)\,(.35) + (12)\,(.25)}{24}$$

c = 0.31.

Finally, compute peak runoff, Q_p, using Equation 4-2:

$$Q_p = Aci$$

$$Q_p = (24)\,(.31)\,(2.65)$$

$Q_p = 20.$ cfs.   (Answer)

Example 4-4

Problem

Suppose a lawn drain is designed for a location in Orange County, California, as shown in Figure 4-15. Compute peak runoff, Q_p, using the Rational Method for a 15-year storm using the proposed lawn drain as the point of analysis. The grate elevation is 377.0.

Solution

First, for an actual design a site visit is essential. However, in this case we will do our best using the aerial topography shown in Figure 4-15.

Begin by delineating the drainage area as depicted in Figure 4-16. Notice that the basin divide cuts through the dwelling at the southerly end of the basin. The divide was drawn by following the contour lines and ignoring the dwelling. If a site visit reveals a different drainage pattern at the house the delineation should be altered.

Next, the drainage basin area is measured by planimeter to be 2.10 acres. At the same time, areas of the various surface types are measured. The area of woods is 1.22 acres and the impervious area is 0.14 acres. The remaining area of 0.74 acre is considered grass.

Runoff coefficients for the three surface types are taken from Table 4-1. Therefore, the composite c value is computed as follows:

Impervious — 0.14 acres @ c = 0.90
Grass   — 0.74 acres @ c = 0.30
Woods   — 1.22 acres @ c = 0.20

$$c = \frac{(.14)\ (.90) + (.74)\ (.30) + (1.22)\ (.20)}{2.10} = 0.28.$$

Next, delineate the hydraulic path as shown in Figure 4-16. Starting at the remotest point, runoff flows as overland flow for approximately 160 feet. At this point the 390-foot contour line which is warped into a swale shape indicates the start of shallow concentrated flow. Shallow concentrated flow then continues all the way to point of analysis, a measured distance of 350 ft.

To find the time for overland flow, t_1, first find the slope:

$$slope = \frac{392.5 - 390}{160} \times 100\% = 1.6\%.$$

Time is found from the nomograph in Appendix C-2 using average grass surface:

$$t_1 = 16.0 \text{ min.}$$

To find the time for shallow concentrated flow, t_2, first find the slope:

$$slope = \frac{390 - 377.0}{350} \times 100\% = 3.7\%.$$

Velocity is found from Figure 4-10 using the line marked "unpaved":

$$v = 3.1 \text{ ft/s.}$$

Time is then computed as

$$t_2 = \frac{d}{v} = \frac{350}{3.1} = 113\,s$$

$t_2 = 1.9$ min.

Therefore, time of concentration is

$t_c = t_1 + t_2$

$t_c = 16.0 + 1.9$

$t_c = 17.9$ min.

The time of concentration, 17.9 min, is taken as the duration of rainfall in determining rainfall intensity, i:

Duration = 17.9 min.

Rainfall intensity is found using the I-D-F curves for Orange County, California in Appendix C-3. For a 15-year storm, interpolate between the 10-year and 25-year curves:

$i = 1.9$ in/h.

Finally, compute peak runoff, Q_p, using Equation 4-2:

$Q_p = Aci$

$Q_p = (2.10)\,(.28)\,(1.9)$

$Q_p = 1.1$ cfs.   (Answer)

The primary application of the Rational Method is in estimating peak runoff for small drainage basins—those which are tributary to minor drainage structures such as storm sewers and small drainage swales. The usual basin size is less than 15 acres and the Rational Method should not be used for drainage areas larger than 100–200 acres.

In addition to size of drainage area, some other important limitations to the Rational Method should be understood. First, the drainage basin surface characteristics should be homogeneous. Although, ideally one c-value should predominate the basin, in practical application this is not possible. This is why a composite c is computed. However, grossly disproportional surface types should be avoided if possible. For example, if a 2.0-acre drainage basin consists of 1.9 acres of paved parking and 0.1 acre of woods at one edge, the hydraulic path would undoubtedly run through the wooded area producing a time of concentration too large to be representative of the vast majority of the basin.

Another drainage basin characteristic to be avoided is any significant ponding within the basin which might affect the peak discharge. Although this condition usually does not occur, when it does, another method of computing peak runoff should be considered.

Finally, it should be remembered that the Rational Method is intended for peak runoff computation, not for the entire hydrograph of a rainfall event. Many methods have been devised to stretch the computation into a hydrograph giving the full relationship of runoff to time following the rainfall. Such computational procedures are loosely referred to as the Modified Rational Method. Although valid results can be obtained for small

Figure 4-15 Topographic map of residential area in Orange County, California. *(Map adapted from Robinson Aerial)*

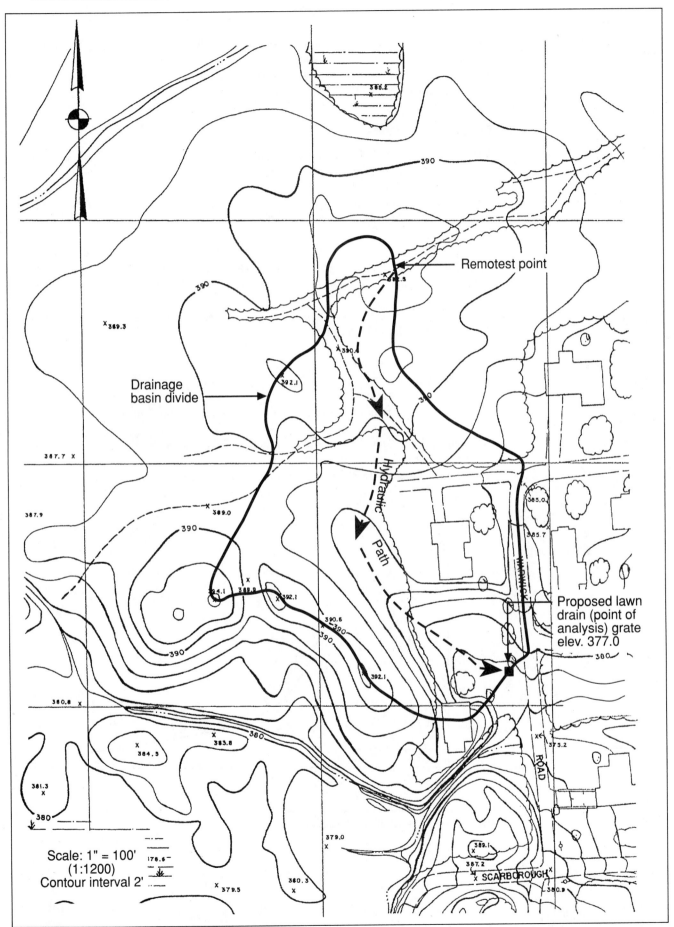

Figure 4-16   Drainage area tributary to proposed lawn drain shown in Figure 4-15. Also shown is hydraulic path. *(Map adapted from Robinson Aerial)*

applications such as minor detention basins, care should be taken when employing the method. Always be aware of the basis of Modified Rational Method you may be using and keep in mind such procedures tend to stretch the Rational Method beyond the task for which it was intended.

4.6   Runoff Hydrographs

Sometimes it is useful to know the entire relationship between runoff and time for a given rainfall event. This relationship, when graphed, is called a *hydrograph* and is shown in general form in Figure 4-17.

A close inspection of Figure 4-17 reveals that, in general, the slope of the rising portion of the hydrograph is steeper than the falling portion. This is characteristic of all hydrographs.

Contrary to the simplified Rational Method model, runoff from actual drainage basins behaves like the hydrograph in Figure 4-17. This is due to a number of factors including the rainfall distribution in time and the surface characteristics of the drainage basin.

Although all runoff hydrographs have the same general shape, they differ in details depending upon the following factors:

1. Amount of rainfall
2. Rainfall pattern
3. Time of concentration
4. Amount of rainfall lost to infiltration, evapotranspiration and surface ponding.

Thus, the peak could be located at various positions along the time axis or the peak could be of various magnitudes or the slope of the rising or falling limbs could be steep or moderate. Figure 4-18 shows the relationship between rainfall and runoff.

Figure 4-17   Typical hydrograph.

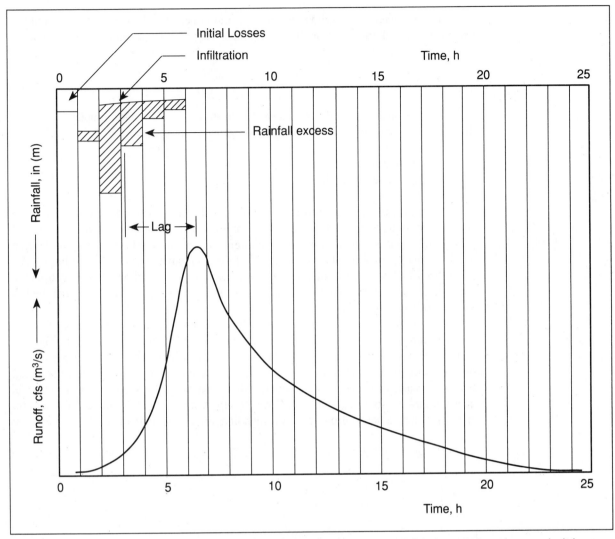

Figure 4-18 Relationship between rainfall and runoff. Although rainfall is inverted on the graph, it is plotted with the same *time* axis as is runoff.

Inspection of Figure 4-18 indicates several factors related to the rainfall/runoff process. First, note that rainfall is plotted in inches at the top of the graph and inverted for convenience. Runoff is plotted in cfs and both plottings share the same time axis. Rainfall is represented as a step graph because it represents the number of inches to fall in each hour. According to this example, rainfall lasted for 6 hours and the total number of inches is the sum of the inches for each of the hours.

Certain portions of the rainfall graph are blank representing *initial losses* (also called initial abstraction) and *infiltration*. The only rainfall to actually run off and reach the point of analysis is the remainder, called *rainfall excess*. The total volume of runoff is the number of inches of rainfall excess multiplied by the area of the drainage basin. This total volume can also be found by computing the area under the hydrograph curve. Initial losses represent the first rainfall to strike the ground and become trapped in depressions or absorbed by plant material. This initial rainfall is completely lost to the runoff process. Infiltration represents that portion of the rainfall which seeps into the ground as runoff flows across the drainage basin. This portion of rainfall also is lost to the runoff process, but is lost at a constant rate throughout the rainfall distribution, not just the initial portion.

The parameter called *lag* is the time separation between the centroid of the rainfall excess graph and the peak of the hydrograph. Lag is related empirically to time of concentration by

$$\text{Lag} = 0.6 t_c. \tag{4-3}$$

Although many methods have been developed to calculate hydrographs for a given drainage basin and storm, they fall into two general categories,

1. Direct measurement hydrograph
2. Synthetic hydrograph

Direct measurement is used for large river basins in which one or more measuring stations are used to record actual hydrographs for every major storm over a number of years. These data, together with related rainfall data, are analyzed statistically to develop a generalized hydrograph that can be applied to any anticipated rainfall in the future. The generalized hydrograph is called a *unit hydrograph* which is defined as the hydrograph resulting from a rainfall excess of one unit. In English units, one unit is 1 inch and in SI units one unit is 1 cm.

The area under the unit hydrograph in Figure 4-19 is equal to the volume of runoff which in turn is equal to one inch multiplied by the area of the drainage basin. The unit hydrograph has a shape characteristic for the drainage basin it represents reflective of the time of concentration and amounts of infiltration, evapotranspiration and surface ponding. It also depends upon the rainfall pattern which can be established for the location in question through statistical analysis. Once the unit hydrograph is established for a drainage basin, any other hydrograph corresponding to any amount of rainfall excess can

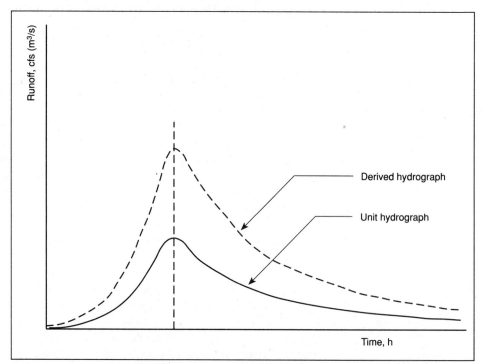

Figure 4-19 Unit hydrograph together with derived hydrograph for the same drainage basin.

easily be computed as a multiple of the unit hydrograph. For example, referring to Figure 4-19, if a rainfall produces a 2-inch rainfall excess, the resulting hydrograph is derived from the unit hydrograph by doubling each ordinate, since 2 inches is double one inch. Doubling each ordinate means at each point along the time axis take the value of runoff on the unit hydrograph and multiply by two to get the value of runoff on the new derived hydrograph. Incidentally, the area under the derived hydrograph is twice that of the unit hydrograph, and the time of the peak runoff is the same for both.

Synthetic hydrographs employ the same concept of unit hydrographs except they are not based upon actual flow measurements. Synthetic hydrographs are used for drainage basins where it is not practical to measure runoff flows, which includes the vast majority of basins. In these cases a method must be devised to reasonably predict the hydrograph resulting from a given rainfall event without ever having experienced actual runoff at the site.

Synthetic hydrograph methodologies developed over the years bear the names of their authors and include the Clark Method, Snyder Method and the SCS Method. The latter is named for the Soil Conservation Service, an agency of the U.S. Department of Agriculture which first published the method in 1972. Although the agency is now called the Natural Resources Conservation Service, NRCS, the hydrograph method is still generally referred to by the original name.

4.7   SCS Method

The SCS Method is a procedure for computing a synthetic hydrograph based upon empirically determined factors developed through research conducted by the Soil Conservation Service. The method is presented in a publication titled *Urban Hydrology for Small Watersheds*, Technical Release No. 55, commonly referred to as TR55.

According to this method, rainfall events are statistically categorized and related to empirically determined unit hydrographs, which in turn yield derived hydrographs as the end product. A variation of the procedure contained in TR55 is also used to compute only peak runoff.

The SCS Method is widely used in the design of major hydraulic structures such as culverts, detention basins, stream relocation, large drainage ditches and the like. These structures generally have tributary drainage areas ranging from a few acres to about 2000 acres. For drainage areas larger than a few square miles, the validity of the SCS Method, or any synthetic hydrograph method, diminishes because the variety of basin characteristics becomes too diverse and because the rainfall pattern does not cover the entire drainage area.

Because TR55, which presents the definitive procedure for the SCS Method, has not yet been converted to SI units, our discussion of the method here will be presented entirely in English units. Any conversion to SI units should be coordinated with SCS to avoid conflicts such as square meters vs hectares for area, centimeters vs meters for rainfall depth, etc.

Before developing the details of the method, let's look at the various parameters involved.

Rainfall—Rainfall is considered as total inches of precipitation, not intensity as in the Rational Method. All rainfall events are assumed to be 24-hour duration. Rainfall patterns have been statistically categorized into four general types depending upon geographical region in the United States.

A map showing geographical distribution of rainfall patterns (called types) is depicted in Appendix D-4. Also, diagrams showing precipitation amounts for various storm return periods are presented in Appendix D-3. These appendices are used when computing runoff by the SCS method.

Infiltration—Loss of rainfall through infiltration, evapotranspiration and minor ponding is described by the curve number, CN, assigned to a drainage basin based upon the type of soil, the pre-existing moisture content of the soil and surface conditions such as flora and impervious structures.

Initial Losses—Initial loss of rainfall, also called initial abstraction, is described by the parameter I_a empirically related to CN.

Time of Concentration—Time of concentration, t_c, is computed as the sum of all travel times along the hydraulic path as in the Rational Method. However, travel time for overland flow is computed by a special empirical formula included in TR55.

Drainage Area—Drainage area, A_m, is the area in square miles enclosed by the basin divide, which is delineated as described in Section 4.2.

To compute runoff by the SCS Method, the first step is to delineate and measure the drainage area tributary to the point of analysis. This is done in the same manner as for the Rational Method.

Next, determine the curve number, CN, for the drainage basin. If the basin consists of a variety of CN's, which is nearly always the case, compute a composite CN which is a weighted average of the individual CN values. A table of CN values is shown in Appendix D-1.

In computing CN, first determine the hydrologic soil group that describes the soils in the drainage basin. Generally, this is done by use of an NRCS Soil Survey of the local region, usually the county within which the drainage area is situated. The Soil Survey contains maps that delineate soil types that have been categorized by field reconnaissance and analysis in accordance with many relevant parameters. Each soil type is assigned a hydrologic soil group of A, B, C or D depending upon its characteristics of infiltration and antecedent moisture condition. Hydrologic soil group A represents the highest rate of infiltration (sandy soil) while D represents the lowest rate (clay soil or rock outcrops). A list of selected soil types together with their respective hydrologic soil group designation is shown in Appendix D-2. These should be referred to when computing runoff by the SCS method.

Using the Soil Survey, identify all soil types within the drainage area and list them together with their hydrologic soil group. If one hydrologic soil group predominates, use that value for CN computation purpose. If there is a mix of groups, they must be considered separately in computing a composite or weighted CN. The following examples illustrate these techniques.

Example 4-5

Problem

A 400-acre drainage basin contains the following soils:

Soil Type	Hyd. Soil Group
Boonton Gravely Loam	B
Parker Gravely Sandy Loam	B
Washington Loam	B

Cover conditions within the drainage basin are as follows:

Cover	Area (acres)
Impervious	5.0
Wooded (good cover)	225
Meadow (good condition)	55
Residential ($^1/_4$-acre lots)	115

Compute the composite CN.

Solution
Use hydrologic soil group B, since it predominates. Next, determine CN for each cover condition using Appendix D-1. Create a table as follows:

Cover	Area (acres)	CN	Product
Impervious	5.0	98	490
Wooded	225	55	12,375
Meadow	55	58	3,190
Residential	115	75	8,625
	400		24,680

$$\text{Weighted CN} = \frac{24680}{400} = 61.7. \quad \text{(Answer)}$$

Example 4-6

Problem
A 250-acre drainage basin contains the following soils:

Soil Type	Area (acres)	Hyd. Soil Group
Califon Loam	100	C
Netcong Loam	150	B

Cover conditions are as follows:

Cover	Area (acres)
Residential ($^1/_2$-acre lots)	190
Wooded (fair condition)	25
Open space (fair condition)	35

Compute the composite CN.

Solution
Since hydrologic soil group is mixed, determine the percentage of each type:

B – 60%,
C – 40%.

Next, determine CN for each cover condition by interpolating between B and C values using Appendix D-1:

| Cover | CN | | Interpolated CN |
	B (60%)	C (40%)	
Residential	70	80	74.0
Wooded	65	76	69.4
Open Space	69	79	73.0

Finally, create a table as follows:

Cover	Area (acres)	CN	Product
Residential	190	74.0	14,060
Wooded	25	69.4	1,735
Open space	35	73.0	2,555
	250		18,350

$$\text{Weighted CN} = \frac{18,350}{250} = 73.4. \qquad \text{(Answer)}$$

After computing CN, the next step is to find the amount of runoff, Q, which is equal to rainfall excess measured in inches. It is the amount of rainfall available for runoff after subtracting initial losses and infiltration and is computed from the following empirical formula:

$$Q = \frac{(P - I_a)^2}{(P - I_a) + S} \tag{4-4}$$

where Q = runoff, in.,
   P = rainfall, in.,
   S = potential maximum retention after runoff begins, in.,
   I_a = initial abstraction (losses), in.

Note: The symbol Q is used for runoff depth in inches to be consistent with symbols used in TR55. Be sure not to confuse this parameter with Q used for runoff rate in the Rational Method.

Initial abstraction is related directly to CN in Table 4-2. Also, S is related directly to CN by the relation

$$S = \frac{1000}{CN} - 10. \tag{4-5}$$

However, Q can be determined without the use of formulas by referring to Figure 4-20.

Table 4-2
I_a Values for Runoff Curve Numbers
(Courtesy of Soil Conservation Service, Technical Release 55.)

Curve number	I_a (in)	Curve number	I_a (in)
40	3.000	70	0.857
41	2.878	71	0.817
42	2.762	72	0.778
43	2.651	73	0.740
44	2.545	74	0.703
45	2.444	75	0.667
46	2.348	76	0.632
47	2.255	77	0.597
48	2.167	78	0.564
49	2.082	79	0.532
50	2.000	80	0.500
51	1.922	81	0.469
52	1.846	82	0.439
53	1.774	83	0.410
54	1.704	84	0.381
55	1.636	85	0.353
56	1.571	86	0.326
57	1.509	87	0.299
58	1.448	88	0.273
59	1.390	89	0.247
60	1.333	90	0.222
61	1.279	91	0.198
62	1.226	92	0.174
63	1.175	93	0.151
64	1.125	94	0.128
65	1.077	95	0.105
66	1.030	96	0.083
67	0.985	97	0.062
68	0.941	98	0.041
69	0.899		

Example 4-7

Problem

Find the runoff depth, Q, for a drainage basin with CN of 78 using the SCS Method. Design rainfall is 6.5 inches.

Solution

Using Table 4-2, I_a is found to be 0.564 in. Next, using Equation 4-5,

$$S = \frac{1000}{CN} - 10$$

$$S = \frac{1000}{78} - 10$$

$$S = 2.28 \text{ in.}$$

Finally, using Equation 4-4,

$$Q = \frac{(P - I_a)^2}{(P - I_a) + S}$$

$$Q = \frac{(6.5 - 0.564)^2}{(6.5 - 0.564) + 2.82}$$

$$Q = 4.0 \text{ in.} \quad \text{(Answer)}$$

As an alternative, use Figure 4-20. Enter the graph at a rainfall of 6.5 inches and trace upward until you reach CN = 78, which is interpolated between the curves for 75 and 80 respectively. Then trace to the left and read the value of Q to be 4.0 in. (Answer)

Following Q, compute time of concentration, t_c, which is the summation of travel times along the hydraulic path. All times are determined as explained in Section 4.3 except for Overland Flow, which is computed by the following empirical formula:

$$T_t = \frac{0.007 \, (nL)^{0.8}}{(P_2)^{0.5} \, s^{0.4}} \tag{4-6}$$

where T_t = overland travel time, hours,
  n = roughness coefficient (See Table 4-3),
  L = length of flow, ft,
  P_2 = 2-year precipitation, in.,
  s = gradient, ft/ft.

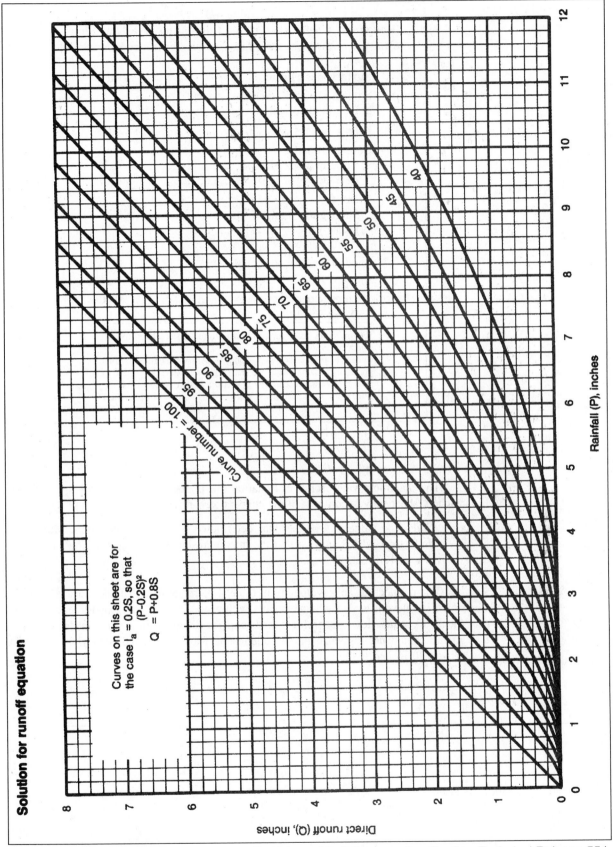

Figure 4-20   Solution of runoff equation. *(Courtesy of Soil Conservation Service, Technical Release 55.)*

Table 4-3
Roughness Coefficients, n, for Computing Overland Flow.
(Courtesy of Soil Conservation Service, Technical Release 55.)

Surface description	n[1]
Smooth surfaces (concrete, asphalt, gravel, or bare soil)	0.011
Fallow (no residue)	0.05
Cultivated soils:	
Residue cover ≤20%	0.06
Residue cover >20%	0.17
Grass:	
Short grass prairie	0.15
Dense grasses[2]	0.24
Bermudagrass	0.41
Range (natural)	0.13
Woods:[3]	
Light underbrush	0.40
Dense underbrush	0.80

[1]The n values are a composite of information compiled by Engman (1986).
[2]Includes species such as weeping lovegrass, bluegrass, buffalo grass, blue grama grass, and native grass mixtures.
[3]When selecting n, consider cover to a height of about 0.1 ft. This is the only part of the plant cover that will obstruct sheet flow.

The maximum value for L in Equation 4-6 is 300 feet. After this length has been reached, sheet flow becomes shallow concentrated flow.

Note: The value of 300 feet was determined by research at NRCS. More recent findings at NRCS suggest that the maximum length of overland flow could be as low as 100 feet.

When selecting an n value from Table 4-3, remember that the surface condition which is relevant is that which prevails within one inch of the ground because that is where the water flows. Therefore, thick underbrush growing several inches above the ground should not influence the selection of n. Most field conditions yield an n-value between 0.15 and 0.40.

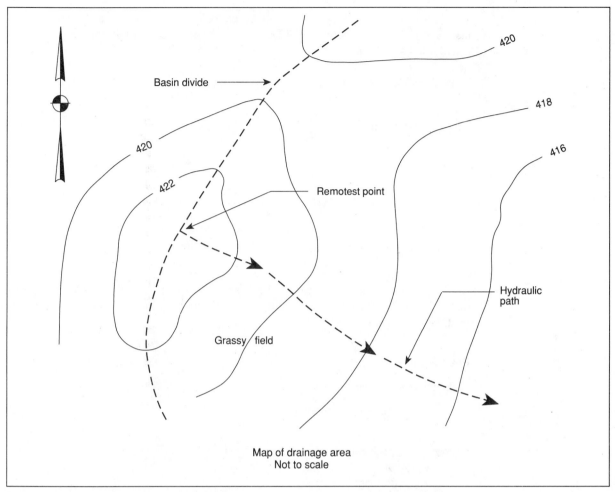

Figure 4-21   Portion of drainage basin showing beginning of hydraulic path.

Example 4-8

Problem
Find the overland travel time T_t for the drainage basin shown in Figure 4-21.

Solution
From inspection of the map shown in Figure 4-21, assume that the length of overland flow is 300 feet. This is because the contour lines do not indicate a swale and the length cannot be greater than 300 feet.

Next, find the slope. Since the elevation of the remotest point is not given, estimate it to be about 423.0. If necessary, sketch a profile of the hydraulic path through the remotest point to estimate the elevation there. From the contour lines, the elevation at the end of 300 feet is 418.0. Therefore, the slope is

$$S = \frac{423.0 - 418.0}{300}$$

$$S = 0.0167 \text{ ft / ft.}$$

Next, estimate the n value. From Table 4-3, n = 0.15.

Two-year precipitation, P_2, is taken from Appendix D-3 (using the 2-year, 24-hour rainfall map) and interpolating between the lines of equal rainfall amount. In this case, since the location is not known, assume $P_2 = 3.3$ inches.

Then, from Equation 4-6

$$T_t' = \frac{0.007 \, (nL)^{0.8}}{(P_2)^{0.5} \, s^{0.4}}$$

$$T_t = \frac{0.007 \, \{(.15)\,(300)\}^{0.8}}{(3.3)^{0.5} \, (.0167)^{0.4}}$$

$$T_t = 0.42 \text{ hour.} \quad \text{(Answer)}$$

Once the drainage area, CN, runoff depth and time of concentration have been determined, it is an easy step to compute either the peak runoff or the entire runoff hydrograph. First, compute peak runoff.

Peak Runoff

Peak runoff is computed by

$$q_p = q_u A_m Q F_p \tag{4-7}$$

where q_p = peak runoff, cfs,
  q_u = unit peak discharge, csm/in,
  A_m = drainage area, mi^2,
  Q = runoff, in.,
  F_p = pond and swamp adjustment factor.

Unit peak discharge, q_u, is another empirical parameter which depends upon rainfall pattern, time of concentration, rainfall and initial abstraction. The unit "csm/in" is an abbreviation meaning cfs per square mile per inch of runoff. Values of q_u can be found by using the graphs shown in Appendix D-5. When using Appendix D-5, be sure to choose the correct rainfall distribution type, which is determined from the map in Appendix D-4.

The pond and swamp adjustment factor is an empirically determined multiplier to account for reduced peak runoff when a small portion of the basin contains swamps or ponds. Values of F_p are listed in Table 4-4. If a large pond or swamp (greater than 5% of the drainage area) is present in the basin, consideration should be given to routing the runoff through the pond or swamp. See Chapter 7 for a discussion of routing.

Table 4-4
Adjustment Factor (F_p) for Pond and Swamp Areas
That Are Spread Throughout the Watershed
(Courtesy of Soil Conservation Service, Technical Release 55.)

Percentage of pond and swamp areas	F_p
0.0	1.00
0.2	0.97
1.0	0.87
3.0	0.75
5.0	0.72

Example 4-9

Problem

Find the peak runoff for a 250-acre drainage basin located in suburban Chicago, Illinois, for a 100-year storm. The basin has the following characteristics:

Soils:	Chancelor	(140 acres)
	Snelling	(110 acres)
Cover:	Residential ($1/2$-acre lots)	(75 acres)
	Wooded (good condition)	(100 acres)
	Meadow	(40 acres)
	Open space (fair condition)	(30 acres)
	Impervious	(5 acres)

Time of concentration: 1.25 hours

Swamp area: 6 acres

Solution

Step 1: Find the area, A_m:

$$A_m = 250 \text{ acres} \times \frac{1 \text{ s.m.}}{640 \text{ acres}} = 0.391 \text{ s.m.}$$

Step 2: Find CN. According to Appendix D-2, the hydrologic soil group for Chancelor soil is C and for Snelling is B. Therefore, CN is determined for each cover condition by interpolating between B and C values using Appendix D-1.

Cover	CN B (44%)	CN C (56%)	Interpolated CN
Residential	75	83	79.48
Wooded	55	70	63.4
Meadow	58	71	65.28
Open Space	69	79	74.6
Impervious	98	98	98.0

Then, create a table as follows:

Cover	Area (acres)	CN	Product
Residential	75	79.48	5961
Wooded	100	63.4	6340
Meadow	40	65.28	2611
Open space	30	74.6	2238
Impervious	5	98.0	490
		Total	17640

$$\text{Weighted CN} = \frac{17640}{250} = 70.6.$$

Step 3: Find the runoff depth, Q. To find Q, first determine the 100-year, 24-hour rainfall for Chicago, Illinois. According to Appendix D-3, P = 5.7 inches. Then, entering 5.7 inches in Figure 4-20 and tracing up to a CN value of 70.6, Q is found to be 2.61 in.

Step 4: Find time of concentration. Time of concentration has already been determined to be 1.25 hours.

Step 5: Determine unit peak discharge, q_u. To find q_u, first determine the rainfall distribution type from Appendix D-4. The basin in question is within the Type II rainfall distribution. Next, compute I_a/P. I_a is found to be 0.833 by interpolating values in Table 4-2. Therefore,

$$\frac{I_a}{P} = \frac{0.833}{5.7} = 0.15.$$

Then, using the unit peak discharge chart for rainfall type II found in Appendix D-5, enter the time of concentration of 1.25 hours and trace upward to the I_a/P value of 0.15 and over to q_u to obtain a value for q_u of 295 csm/in.

Step 6: Determine F_p from Table 4-4. Using a swamp percentage of 2.4%, interpolate between 1.0% and 3.0% to obtain a value for F_p of 0.79.

Step 7: Finally, compute q_p from Equation 4-7.

$$q_p = q_u A_m Q F_p$$

$$q_p = (295) \, (.391) \, (2.61) \, (.79)$$

$$q_p = 238 \text{ cfs.} \qquad \text{(Answer)}$$

Example 4-10

Problem

Compute the peak runoff, q_p, using the SCS Method for a 100-year storm for a watershed located near Laguna Beach, California and having the following parameters:

$$A_m = 0.873 \text{ s.m.}$$

$$CN = 74$$

$$t_c = 1.45 \text{ hours}$$

Solution

Step 1: The watershed area, A_m, is already known.

Step 2: The curve number, CN, also is known.

Step 3: Find the runoff depth, Q. To find Q, first determine the 100-year, 24-hour rainfall for Laguna Beach. According to the map of southern California in Appendix D-3, P = 6.0 inches. Entering 6.0 inches in Figure 4-20 and tracing up to a CN value of 74, Q is found to be 3.15 inches.

Step 4: Time of concentration, t_c, is already known.

Step 5: Determine unit peak discharge, q_u. To determine q_u, first find the rainfall distribution type from Appendix D-4. The watershed in question is within the Type I region. Next, compute I_a/P. I_a is found to be 0.703 from Table 4-2. Therefore,

$$\frac{I_a}{P} = \frac{.703}{6.0} = 0.12.$$

Then, using the unit peak discharge chart for rainfall type I found in Appendix D-5, enter the time of concentration of 1.45 hours and trace upward to the I_a/P value of 0.12 and over to q_u to obtain a value for q_u of 160 csm/in.

Step 6: The value of F_p is 1.0 since no swamp areas or ponds are present in the watershed.

Step 7: Finally, compute q_p from Equation 4-7:

$$q_p = q_u A_m Q F_p$$

$$q_p = (160)\ (.873)\ (3.15)\ (1)$$

$$q_p = 440 \text{ cfs.} \qquad \text{(Answer)}$$

Runoff Hydrograph

Now, compute the entire runoff hydrograph using the SCS method. This procedure is used for complex watersheds which must be divided into sub-areas or for detention basin routing computations. Appendix D-6 contains a selected series of unit hydrographs which apply to various drainage basin characteristics. The numbers in each unit hydrograph are unit discharges, q_t, expressed in csm/in. Each q_t value corresponds to a time, t, representing the elapsed time from the beginning of the storm. To determine the desired hydrograph, first select the appropriate unit hydrograph from Appendix D-6 by using the following parameters:

1. Time of concentration, t_c
2. I_a/P
3. Rainfall distribution type
4. Travel time, T_t

Travel time is used when the point of analysis is some distance downstream of the actual drainage basin and will be explained later. For the usual point of analysis we are discussing here, travel time is zero. Each page or chart of Appendix D-6 contains 36 unit hydrographs, one for each line. All the unit hydrographs on each page represent one rainfall type and one time of concentration. So, for instance, Chart 1 is for rainfall Type II and $t_c = 0.4$ h. On each chart, the unit hydrographs are divided into three groups, each one representing a different value of I_a/P. The column of numbers on the left side contain values of travel time, T_t. Therefore, by knowing the rainfall distribution together with t_c, I_a/P and T_t, you can locate the specific line in Appendix D-6 representing the desired unit hydrograph. To complete the unit hydrograph, each value in the line is paired with one of the t-values listed across the top of the page.

After selecting the unit hydrograph, compute the derived hydrograph by multiplying each unit discharge by the amount A_mQ. Therefore,

$$q = q_t A_m Q \qquad\qquad (4\text{-}8)$$

where q = derived hydrograph ordinate at time t, cfs,
$\quad\quad\quad q_t$ = unit discharge at time t, csm/in,
$\quad\quad\quad A_m$ = drainage area, s.m.,
$\quad\quad\quad Q$ = runoff depth, in.

It should be noted that Equation 4-8 does not include a pond factor term as does Equation 4-7. This is because ponds and swamps are handled differently using the hydrograph procedure. If a significant pond or swamp is present in the watershed, it should be isolated in a sub-area and then either routed or accounted for by adjusted CN and t_c values. Routing is a mathematical procedure which will be explained in Chapter 7.

Example 4-11

Problem
Compute the 100-year runoff hydrograph for a 190-acre watershed located near Nashville, Tennessee having a CN value of 78 and a time of concentration of 1.05 hours.

Solution
Step 1: Find area, A_m:

$$A_m = 190 \text{ acres} \times \frac{1 \text{ s.m.}}{640 \text{ acres}} = 0.2969 \text{ s.m.}$$

Step 2: The curve number, CN, is known.

Step 3: Find runoff depth, Q, knowing CN = 78. To find Q, first determine the 100-year, 24-hour rainfall for central Tennessee. According to Appendix D-3, P = 6.8 inches. Then, entering 6.8 inches in Figure 4-20 and tracing up to a CN value of 78, Q is found to be 4.3 inches.

Step 4: Time of concentration has already been determined to be 1.05 hours.

Step 5: Locate the appropriate unit hydrograph in Appendix D-6 using the following values

  1. t_c = 1.05 h
  2. I_a/P = .564/6.8 = 0.08
  3. Rainfall distribution Type II (from Appendix D-4)
  4. T_t = 0.

The unit hydrograph is found on page 375, line 1 of Appendix D-6 and is reproduced in Table 4-5 as the first two columns. Note that when selecting the appropriate unit hydrograph, if any of the parameters t_c, I_a/P or T_t do not match those in Appendix D-6, you must round off to the nearest value. In this case, t_c = 1.05 h was rounded to t_c = 1.0 h and I_a/P = 0.08 was rounded to I_a/P = 0.10.

**Table 4-5
Runoff Hydrograph for Example 4-10.**

Hydrograph Time (h)	Unit Discharges (csm/in)	Hydrograph Ordinates (cfs)
11.0	11	14
11.3	15	19
11.6	20	26
11.9	29	37
12.0	35	45
12.1	47	60
12.2	72	92
12.3	112	143
12.4	168	214
12.5	231	295
12.6	289	369
12.7	329	420
12.8	357	456
13.0	313	400
13.2	239	305
13.4	175	223
13.6	133	170
13.8	103	131
14.0	83	106
14.3	63	80
14.6	50	64
15.0	40	51
15.5	33	42
16.0	29	37
16.5	26	33
17.0	23	29
17.5	21	27
18.0	20	26
19.0	17	22
20.0	15	19

Step 6: Using Equation 4-8, compute each q value:

$q = q_t A_m Q$
$q = q_t (.2969) (4.3)$
$q = 1.277 q_t.$

Each q value becomes a value in the third column in Table 4-5, which is the desired hydrograph and answer to the problem.

Note that the peak discharge, according to the hydrograph, is 456 cfs occurring at time 12.8 hours.

Sub-Basins

When using the SCS Method, as with any other runoff computation, if the delineated drainage basin is not sufficiently homogeneous, it must be subdivided into smaller units called *sub-basins*. We must remember that one of the underlying assumptions of the SCS Method is a drainage basin with relatively uniform characteristics throughout. These characteristics include cover conditions, average slopes and time of concentration.

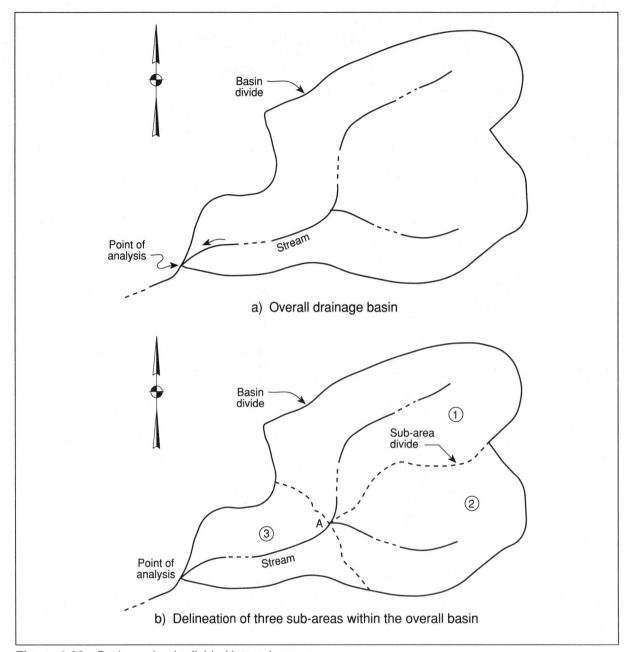

a) Overall drainage basin

b) Delineation of three sub-areas within the overall basin

Figure 4-22   Drainage basin divided into sub-areas.

Figure 4-22(a) shows the outline of a drainage basin with a stream branched into two areas with differing characteristics. Figure 4-22(b) shows how the basin can be subdivided into three sub-areas for computation purposes. Sub-areas 1 and 2 start at point A which is the *confluence* of the northerly and southerly branches of the stream.

The hydrograph for this basin at the point of analysis is the sum of the hydrographs generated by each of the three sub-areas. Each sub-area produces a hydrograph just like any single watershed does.

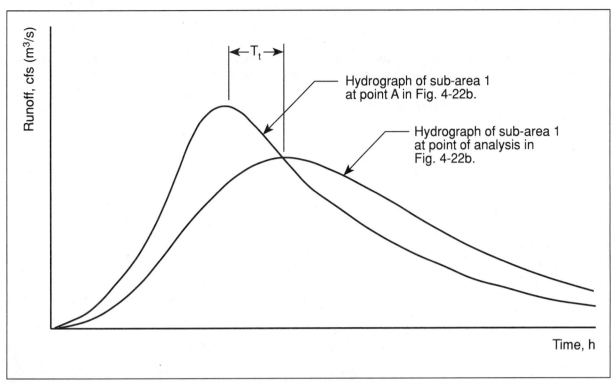

Figure 4-23   Attentuation of hydrograph due to travel in stream.

Sub-area 1 in Figure 4-22(b) produces a hydrograph at point A and so does Sub-area 2. However, we are not interested in the hydrograph at point A; we want to compute the hydrograph at the point of analysis. This is where travel time, T_t, becomes important. As the runoff from Sub-area 1 flows down the stream toward the point of analysis, its hydrograph changes. The mere act of flowing down the stream alters the time relationship of the runoff. This alteration is called *attenuation* and is illustrated in Figure 4-23.

Figure 4-23 shows that the attenuated hydrograph is squatter and longer than that at point A. Attenuation causes the magnitude of the peak discharge to diminish and the time of the peak to move to the right by an amount equal to the travel time T_t. However, the areas under the two hydrographs are equal since the total amount of runoff has not changed.

It is important to remember in this discussion of attenuation that the hydrographs depicted in Figure 4-23 represent only the runoff from Sub-area 1, even though this runoff becomes mixed with water from the other two sub-areas as it moves down the stream.

So to compute the total runoff hydrograph for the drainage area in Figure 4-22, compute three hydrographs as they would appear at the point of analysis and add them together. Hydrographs are added using the principle of *superposition* in which each ordinate of the resulting hydrograph is the sum of three ordinates, one from each hydrograph, for each time value along the hydrograph. The following example illustrates the computation of a hydrograph where sub-areas are involved.

Example 4-12

Problem

Find the runoff hydrograph for the drainage basin depicted in Figure 4-22 for a 100-year storm. Assume the basin is located in central Ohio and has the following parameters:

Sub-Area 1: A_m = 0.250 s.m.
CN = 75
t_c = 1.0 hours

Sub-Area 2: A_m = 0.190 s.m.
CN = 60
t_c = 1.25 hours

Sub-Area 3: A_m = 0.100 s.m.
CN = 70
t_c = 0.50 hours

Travel time from Point A to Point of Analysis: 0.20 hours

Solution

First, find the total rainfall, P, from Appendix D-3:

P = 5.0 in.

Next, find the rainfall distribution type from Appendix D-4:

Type II.

Now, determine the appropriate unit hydrographs for Sub-areas 1, 2 and 3 and use them to compute the three derived hydrographs at the point of analysis. Unit hydrographs are found using the following parameters:

Sub-Area	t_c (h)	I_a/P	T_t (h)
1	1.0	$\dfrac{.667}{5.0} = 0.13$	0.20
2	1.25	$\dfrac{1.333}{5.0} = 0.27$	0.20
3	0.50	$\dfrac{.857}{5.0} = 0.17$	0.00

Unit hydrographs are found in Appendix D-6 and are reproduced in Table 4-6 as columns 2, 4 and 6. Values of derived hydrograph ordinates are listed in columns 3, 5 and 7 and are computed by multiplying by the amount A_mQ. Values of Q are found from Equations 4-4 and 4-5 or from Figure 4-20:

Sub-Area	A_m (s.m.)	Q (in)	$A_m Q$ (s.m.-in)
1	.250	2.45	0.6125
2	.190	1.30	0.2470
3	.100	2.04	0.2040

The answer to the problem is the hydrograph shown in Column 8 of Table 4-6 which is computed by adding values in Columns 3, 5 and 7 for each time value in Column 1. Graphical depiction of the hydrograph is shown in Figure 4-24.

Careful scrutiny of Table 4-6 reveals an important characteristic of working with sub-areas. Peak discharges at the point of analysis for the three sub-areas are 203 cfs, 59 cfs and 108 cfs, respectively. However, the total peak discharge is 282 cfs which is less than the sum of the three individual peaks. This is due to the timing of the peaks. Since they do not occur at the same time, they are not added directly when computing the superposition of the three hydrographs.

Table 4-6
Runoff Hydrograph for Example 4-11.

(1) Hydrograph Time (h)	(2) Sub-Area 1 Unit Discharges (csm/in)	(3) Sub-Area 1 Hydrograph Ordinates (cfs)	(4) Sub-Area 2 Unit Discharges (csm/in)	(5) Sub-Area 2 Hydrograph Ordinates (cfs)	(6) Sub-Area 3 Unit Discharges (csm/in)	(7) Sub-Area 3 Hydrograph Ordinates (cfs)	(8) Hydrograph (cfs)
11.0	10.	6.1	0	0	17	3.5	9.6
11.3	13	8.0	0	0	23	4.7	13
11.6	17	10.	0	0	32	6.5	17
11.9	23	14	0	0	57	12	26
12.0	26	16	0	0	94	19	35
12.1	30.	18	0	0	170	35	53
12.2	38	23	1	0.2	308	63	86
12.3	54	33	4	1.0	467	95	129
12.4	82	50.	14	3.5	529	108	162
12.5	123	75	31	7.7	507	103	186
12.6	176	108	58	14	402	82	204
12.7	232	142	93	23	297	61	226
12.8	281	172	133	33	226	46	251
13.0	332	203	202	50.	140	29	282
13.2	303	186	239	59	96	20.	265
13.4	238	146	231	57	74	15	218
13.6	179	110	199	49	61	12	171
13.8	136	83	165	41	53	11	135
14.0	105	64	138	34	47	10.	108
14.3	76	47	108	27	41	8.4	82
14.6	59	36	87	21	36	7.3	64
15.0	45	28	68	17	32	6.5	52
15.5	35	21	55	14	29	5.9	41
16.0	30.	18	47	12	26	5.3	35
16.5	27	17	41	10.	23	4.7	32
17.0	24	15	37	9.1	21	4.3	28
17.5	22	13	33	8.2	20.	4.1	25
18.0	20.	12	31	7.7	19	3.9	24
19.0	18	11	28	6.9	16	3.3	21
20.0	16	10.	25	6.2	14	2.9	19

Figure 4-24 Runoff hydrographs for Example 4-11.

Limitations of the hydrograph method discussed here include the need to perform time consuming computations and the loss of precision due to the rounding off of t_c, I_a/P and T_t values when selecting unit hydrographs. To overcome both problems, the use of computer software is recommended. The Soil Conservation Service has published the program TR-20 which enables the computing of SCS hydrographs on a P.C. Several other programs are available as well. Refer to Appendix E for a list of applicable software.

PROBLEMS

1. Carefully trace the delineated drainage area shown in Figure 4-25 onto tracing paper and then determine the drainage area in a) acres and b) square meters.

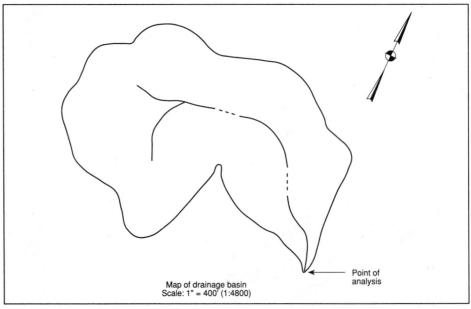

Map of drainage basin
Scale: 1" = 400' (1:4800)

Point of analysis

Figure 4-25 Delineation of drainage basin.

Side Street

Grass

Woods

Basin divide

Point of analysis
(inlet)

Main street

Map of drainage basin
Scale: 1" = 100' (1:1200)

Figure 4-26   Delineation of drainage basin.

2. Trace the delineated drainage area shown in Figure 4-26 onto tracing paper and then
   determine a) drainage area in 1) acres and 2) square meters b) composite runoff
   coefficient (Rational Method).

3. Find the composite runoff coefficient, c, for a drainage basin having the following
   cover conditions:

Pavement	2500	s.f.
Roofs	2000	s.f.
Driveway	800	s.f.
Grassed areas	0.45	acres
Woods	1.21	acres

4. Find the composite runoff coefficient, c, for a drainage basin having the following
   cover conditions:

Roof	700	m^2
Pavement	2400	m^2
Lawns, sandy soil Average 5% slope	1.50	hectares
Unimproved	2.75	hectares

5. Determine time of concentration, t_c, for use in the Rational Method for a drainage basin having a hydraulic path described as follows:

   Overland flow:
      average grass, 240' long, 2.5% average slope
   Shallow concentrated flow:
      450' long, 4.8% average slope
   Gutter flow:
      200' long, 1.9% average slope

6. Determine time of concentration, t_c, for use in the SCS Method for a drainage basin located near Bismark, North Dakota, having a hydraulic path described as follows:

   Overland flow:
      average grass, 300' long, 1.6% average slope
   Shallow concentrated flow:
      680' long, 3.2% average slope
   Stream flow:
      2950' long, 0.61% average slope
   Average cross section:

7. A drainage basin located in New Jersey has a time of concentration of 18 minutes. Find the rainfall intensity, i, for use in the Rational Method for a 25-year storm.

8. A drainage basin located in the Atlanta, Georgia area has a time of concentration of 2.5 hours. Find the rainfall intensity, i, for use in the Rational Method for a 10-year storm.

9. Compute the peak runoff, Q_p, for a 15-year storm using the Rational Method for a drainage basin located in Phoenix, Arizona and having parameters as follows:

   A = 14.06 acres
   Time of concentration:
      Overland flow:
        average grass, 300' long
        5.0% average slope
      Shallow concentrated flow:
        50' long, 7.0% average slope
   Runoff coefficient:
        Impervious: 0.06 acres
        Grass: 6.5 acres
        Woods: 7.5 acres

   Note: In the Arizona I-D-F curves, a 15-year storm is located by the corresponding P_1 value which for Phoenix is 1.75 inches.
   Express Q_p in a) cfs and b) m³/s.

10. Compute the peak runoff, Q_p, for the drainage area in Problem 9 for a 50-year storm. Express Q_p in a) cfs and b) m³/s.

   Note: In the Arizona I-D-F curves, the P_1 value corresponding to a 50-year storm is 2.25 inches.

11. Compute the composite curve number, CN, for a drainage basin having the following soils and cover conditions:

      Soils: Bartley Loam: 45 acres
                 Parker Loam: 25 acres

      Cover: Impervious   2.0 acres
                 Residential ($^1/_3$-acre lots)   20 acres
                 Wooded (fair cover)   28 acres
                 Grass (fair condition)   20 acres

12. Compute the composite curve number, CN, for a drainage basin having the following soils and cover conditions:

      Soils: Maplecrest   120 acres
                 Hamel   32 acres
                 Vinsad   55 acres
                 Alluvial land   2 acres

      Cover: Impervious   1.0 acre
                 Wooded   100 acres
                 Disturbed soil (bare)   23 acres
                 Brush (poor condition)   85 acres

13. A drainage basin located in the Chicago, Illinois, area has a composite CN value of 71.5. Compute the runoff depth, Q, for a 100-year storm.

14. A drainage basin located near Columbus, in central Ohio, has a composite CN value of 66. Compute the runoff depth, Q, for a 25-year storm.

15. Compute the peak runoff, q_p, using the SCS Method for a 100-year storm for a drainage basin located in the Atlanta, Georgia, area and having the following parameters:

      A_m = 1.413 s.m.
      CN = 68.0
      t_c = 2.35 hours

16. Compute the peak runoff, q_p, using the SCS Method for a 50-year storm for a drainage basin located at the western- most end of the Florida panhandle and having the following parameters:

      A_m = 1.250 s.m.
      CN = 61.2
      t_c = 2.75 hours

In Problems 17 through 22, measure the appropriate drainage area by tracing the delineation onto tracing paper. Measurement can be done by planimeter or by approximating geometric shapes.

17. Measure the area of the drainage basin shown in Figure 4-27. Express the area in a) acres and b) m².

18. Measure the area of the drainage basin shown in Figure 4-28. Express the area in a) acres and b) m².

19. Measure the area of the drainage basin shown in Figure 4-29. Express the area in a) acres and b) m².

20. Measure the area of the drainage basin shown in Figure 4-30. Express the area in square miles.

21. Measure the area of the drainage basin shown in Figure 4-31. Express the area in square miles.

22. Measure the area of the drainage basin shown in Figure 4-32. Express the area in square miles.

Figure 4-27 Delineation of drainage basin located in New Jersey. *(Map adapted from Robinson Aerial.)*

Figure 4-28 Delineation of drainage basin located in Orange County, California. *(Map adapted from Aero Service.)*

Figure 4-29 Delineation of drainage basin located in Pennsylvania, Region 1. *(Map adapted from Aero Service.)*

Figure 4-30 Delineation of drainage basin located in San Diego, California. *(Map adapted from Aero Service.)*

Figure 4-31 Delineation of watershed located in western Pennsylvania. *(Courtesy of U.S. Geological Survey.)*

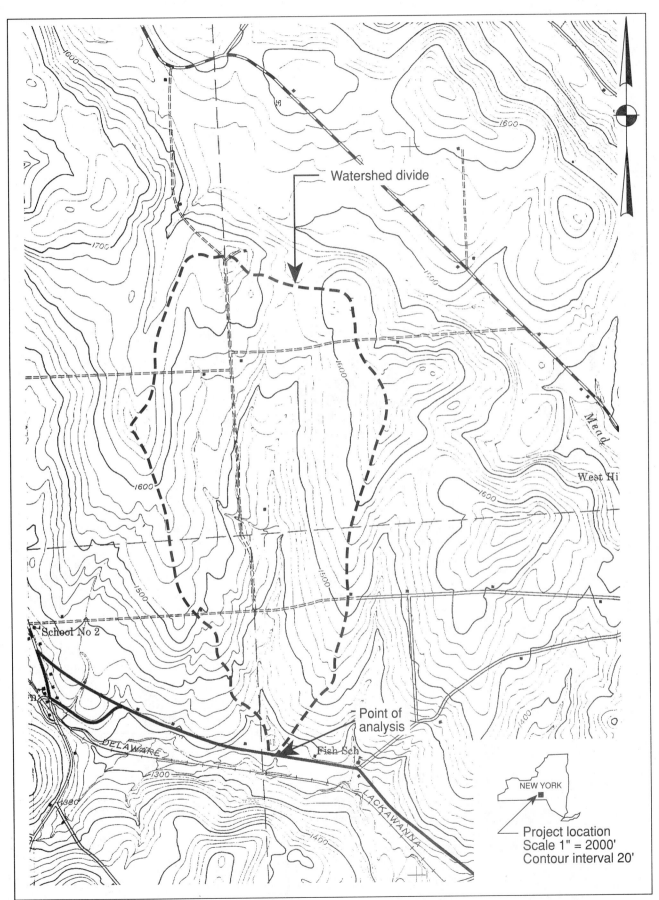

Figure 4-32 Delineation of watershed located in central New York. *(Courtesy of U.S. Geological Survey.)*

23. Compute the runoff hydrograph for a 50-year storm for a drainage basin located in Tulsa, Oklahoma, having the following parameters:

    A_m = 0.892 s.m.
    CN = 76.0
    t_c = 1.45 hours

24. Compute the runoff hydrograph for a 100-year storm for a drainage basin located in the Albany, New York, area (shaped like that shown in Figure 4-20) and having the following parameters:

    Sub-Area 1: A_m = 1.012 s.m.
    CN= 82
    t_c = 1.0 hours

    Sub-Area 2: A_m = 0.761 s.m.
    CN= 63
    t_c = 2.0 hours

    Sub-Area 3: A_m = 0.550 s.m.
    CN= 71
    t_c = 0.60 hours

    Travel time from Point A to Point of Analysis: 0.26 hours.

In Problems 25 through 30, make your best assumptions based upon the available maps. (Normally, a site visit is needed to confirm field conditions.)

25. Calculate peak discharge, Q_p, for the drainage basin shown in Figure 4-27 for a 15-year storm using the Rational Method. The basin is located in New Jersey. For an average stream cross-section, use that in Example 4-3. Assume the entire basin is wooded.

26. Calculate peak discharge, Q_p, for the drainage basin shown in Figure 4-28 for a 25-year storm using the Rational Method. The basin is located in Orange County, California.

27. Calculate peak discharge, Q_p, for the drainage basin shown in Figure 4-29 for a 5-year storm using the Rational Method. The basin is located in Pennsylvania, Region 1.

28. Calculate peak discharge, q_p, for the watershed shown in Figure 4-30 for a 100-year storm using the SCS Method. The basin is located in San Diego, California. For an average stream cross-section, use that in Problem 6. Assume the entire watershed is wooded, hydrologic soil group B. Two year rainfall, P_2, is 1.8 inches.

29. Calculate peak discharge, q_p, for the watershed shown in Figure 4-31 for a 25-year storm using the SCS Method. The basin is located in western Pennsylvania. For an average stream cross-section, use that in Problem 6. Assume hydrologic soil group C.

30. Calculate peak discharge, q_p, for the watershed shown in Figure 4-32 for a 50-year storm using the SCS Method. The basin is located in central New York. For an average cross-section, use that in Problem 6. Assume hydrologic soil group B (65%), C (25%), D (10%).

FURTHER READING

Akan, A. O. (1993). *Urban Stormwater Hydrology, A Guide to Engineering Calculations.* Lancaster, PA: Technomic Publishing.

Chow, V. T., Maidment, D. R. and Mays, L. R. (1988). *Applied Hydrology.* New York: McGraw-Hill.

Commonwealth of Pennsylvania, PENNDOT, Bureau of Design. (1990). *Design Manual, Part 2, Highway Design*, Publication 13. Harrisburg, PA: Commonwealth of Pennsylvania.

Hershfield, D. M. (1961). *Rainfall Frequency Atlas of the United States for Durations from 30 Minutes to 24 hours and Return Periods of 1 to 100 Years*, Technical Paper No. 40. Silver Springs, MD: U.S. National Weather Service.

McCuen, R. H. (1982). *A Guide to Hydrologic Analysis Using SCS Methods*. Englewood Cliffs, N.J.: Prentice-Hall.

New Jersey Department of Transportation, Division of Roadway Design, Bureau of Roadway Design Standards. (1994). *Roadway Design Manual*. Trenton, N.J.: Department of Transportation.

Ponce, V. M. (1989). *Engineering Hydrology: Principles and Practices*. Englewood Cliffs, N.J.: Prentice Hall.

Urban Water Resources Research Council of ASCE and the Water Environment Federation. (1992). *Design and Construction of Urban Stormwater Management Systems*. New York, NY and Alexandria, VA: ASCE.

U. S. Soil Conservation Service. (1972). *National Engineering Handbook*. Springfield, VA: U.S. Department of Agriculture.

U. S. Soil Conservation Service. (1986). *Urban Hydrology for Small Watersheds*. Technical Release 55. Springfield, VA: U. S. Department of Agriculture.

Viessman, W., Jr., Lewis, G. L. and Knapp, J. W. (1989). *Introduction to Hydrology*. (3rd ed.). New York: Harper and Row.

Wanielista, M. P. (1990). *Hydrology and Water Quality Control*. New York: John Wiley & Sons.

Wanielista, M. P., Yousef, Y. A. (1993). *Stormwater Management*. New York: John Wiley & Sons.

5

STORM SEWER DESIGN

Storm sewers are underground pipes used to convey storm water from developed areas safely and conveniently into natural bodies of water such as streams and lakes. They are used typically in roads, parking areas and sometimes in lawns.

Prior to the development of storm sewers, urban planners managed stormwater by channeling it in a series of swales along streets and alleys and eventually into streams, a practice still in use today in underdeveloped countries. Because the water remained on the ground surface, inconvenience and disease transmission resulted. The introduction of storm sewers allows for the development of land with modern advancements in transportation virtually unencumbered by stormwater problems.

Throughout the past century, storm sewers have been used in urban settings to convey sewage waste as well as storm water in the same pipes. Such systems are called *combined sewers* and have now been nearly eliminated in favor of separate storm sewers and sanitary sewers.

In this chapter, we will learn the basic principles in the hydraulic design of storm sewers. Of course, the design of a storm sewer includes many more factors than hydraulics, such as structural stability, construction methods, cost, etc. Although some of these factors will be mentioned, the principle emphasis will be on hydraulic design.

OBJECTIVES

After completing this chapter, the reader should be able to:
- Layout a storm sewer system in a road or parking area
- Interpret a storm sewer profile
- Delineate incremental drainage areas in a standard storm sewer design
- Compute pipe sizes in a standard storm sewer design
- Design riprap outfall protection for a storm sewer outlet
- Relate a standard storm sewer design to an actual case study

5.1 FUNDAMENTAL CONCEPTS

The three principle components of a storm sewer system are the inlet, the pipe and the outflow headwall. The *inlet* is a structure designed to allow stormwater into the system; the *pipe* conveys stormwater toward a receiving stream; and the *headwall* allows stormwater to exit the system. Figure 5-1 depicts these basic elements of a storm sewer system.

Figure 5-1   Essential elements of a storm sewer system.

Inlets usually are precast concrete square structures placed in the ground and topped with a cast iron grate at ground level. Inlets are often mistakenly called *catch basins* which are similar storm sewer structures with a slightly different purpose. Figure 5-2 depicts a typical inlet. Catch basins are designed with a bottom set lower than the incoming and outgoing pipes to provide a sump which acts as a sediment trap.

In addition to allowing stormwater into the system, inlets provide access to the pipes for maintenance as well as a point for the pipes to change direction since pipes must normally be laid straight without curves.

Figure 5-2   Construction detail of typical inlet.

Pipes are manufactured in a variety of materials including concrete, reinforced concrete, corrugated metal and plastic. In general, the cross-sections of pipes are round or elliptical, with most pipes being round. Figure 5-3 depicts some typical examples. Pipes are generally laid in trenches in accurate straight alignments and grades. Pipe sections are designed to fit together as the pipe is laid from its downstream end to its upstream end.

Figure 5-3 Typical storm sewer pipe sections.

The most fundamental element of a storm sewer design is choosing a pipe with sufficient size and grade to convey the design runoff from one inlet to the next. This is accomplished by comparing the capacity in cfs (m³/s) of the pipe to the design flow in cfs (m³/s). If the capacity is greater than the flow, the pipe is adequate for use in the storm sewer system.

However, several other factors must also be considered in designing a pipe segment. First, the velocity of flow must be properly controlled. On the one hand, velocity should not be allowed to be too low or silt and debris may be deposited along the pipe, but on the other hand, excessive velocity should be avoided. The minimum velocity needed to avoid siltation is usually called the *cleansing velocity* and generally is taken as 2.0 ft/s (0.60 m/s). Therefore, the pipe should be designed so that when flowing half full a minimum velocity of 2.0 ft/s (0.60 m/s) is maintained.

Excessive velocity should be avoided because the result could be damage to the pipe or surcharging of an inlet at a bend point in the system. The choice of a maximum velocity is somewhat elusive because it depends upon site specific conditions. Some municipal ordinances set a design standard of 10 ft/s (3.0 m/s), and in such cases this would become the maximum velocity. In the absence of local regulation, 15 ft/s (4.6 m/s) is a reasonable value for design purposes.

Note: An inlet becomes surcharged when it fills with water due to the inability of the downstream pipe to adequately convey the flow.

In order to maintain sufficient velocity and to keep the storm sewer system flowing smoothly, head losses within the system should be kept to a minimum. This is accomplished principally by streamlining the transitions from one pipe segment to the next.

A few basic rules are generally used in currently accepted engineering practice and are also found in some municipal regulations. First, choose a minimum pipe diameter of 12 inches. (Some municipal regulations require a 15-inch minimum.) This helps prevent clogging and also facilitates maintenance. Second, as you choose pipe sizes throughout a storm sewer system, each pipe segment must be equal to or larger than the segment immediately upstream. This is another measure to prevent clogging. So, for example, if one pipe segment has a diameter of 24 inches and the next downstream segment need only be an 18-inch pipe to convey the design stormwater flow, you must choose a 24-inch pipe size nonetheless. A third rule relates to transitions from one size pipe to a larger size. When making a transition at an inlet, the vertical alignment of the incoming and outgoing pipes should be such that the crowns line up (not the inverts). This promotes smooth flow and helps prevent a backwater in the upstream pipe. Figure 5-4 illustrates these general rules.

5.2   DESIGN INVESTIGATION

Before beginning a storm sewer design, certain essential data must be acquired through field investigations and other sources. These can be summarized in general as follows:

1. Project meeting
2. Topographic map
3. Site reconnaissance
4. Local Land Development ordinance
5. Related engineering designs or reports

First, a project meeting must be held with the developer, that is, the person who will be paying for the new storm sewer system. For example, the developer could be an entrepreneur who wishes to build a building for profit or a municipality which wishes to improve the drainage on one of its roads.

At the project meeting, you should determine the scope of the project as clearly as possible. Project scope includes the physical boundaries of the site, the nature of the development and whether any off-site improvements will be required. The developer should provide a boundary survey of his land and a sketch showing the proposed project

a) Use a minimum pipe size of 12" diameter

PROFILE
b) Never decrease pipe sizes going downstream

SECTION
c) When transitioning from smaller to larger pipe, match the crown alignments (not the inverts).

Figure 5-4   Some basic rules in storm sewer design.

in schematic form. If he has no survey, it should be determined at the meeting whether or not a boundary survey must be conducted. Also, it should be determined whether or not a topographic survey must be conducted for the project.

Boundary and topographic surveys, if required, are part of the basic engineering services to be provided for the project.

A topographic survey is almost always needed in order to delineate drainage areas for runoff computation. For most designs, watersheds are relatively small requiring detailed topography to adequately determine drainage areas. Typically, maps are prepared at a scale of 1"= 50' (1:600) or larger with contour lines drawn at a 2-foot interval. USGS quadrangle maps drawn at a scale of 1"= 2000' (1:24000) and contour interval of 10 or 20 feet are not sufficiently detailed for storm sewer design.

A useful topographic map can be prepared by one of three usual methods:

1. The map can be based upon a ground topographic survey, that is, one conducted by a survey crew on the site.
2. The map can be based upon an aerial topographic survey which is drawn from aerial photographs and calibrated by limited ground survey control.
3. A combination of aerial and ground survey can be used.

If the storm sewer system is to be part of a new development where all roads and parking areas will be designed from scratch, aerial topography is adequate. However, if existing roads are involved and if the proposed storm sewer must connect to existing pipes, then more precise topography is needed to specify the existing features. In such cases, a combination of aerial and ground topography should be used.

After the topographic map has been obtained, a site visit must be conducted to verify the information on the map and to determine certain key information not shown. For instance, the type of ground cover should be noted to help in choosing runoff coefficients. Also, if an existing road has a functioning gutter, it should be noted in order to help determine the hydraulic path. In addition, if an existing building has a complex drainage pattern not evident in the topographic map, this condition should be noted.

Another purpose of the site visit is to note any underground utilities, including storm sewers, not shown on the topographic map. Any such missing utilities should then be added to the map. Applicable local ordinances must be obtained as part of the design investigation. Local ordinances often specify such design parameters as design storm, runoff coefficients or inlet type and spacing. If any county or state regulations apply, these must be considered as well. Also, the storm intensity-duration-frequency (I-D-F) curves for the project location must be obtained. Finally, if any previous development in the project vicinity has generated design plans or reports, these should be acquired. Existing plans may have a bearing on the proposed project and may also indicate which ideas have already been approved by the local approving agency.

It is also often helpful to schedule a preliminary meeting with the municipal engineer who has jurisdiction over the project. Such a meeting will establish a rapport with the engineer and may reveal problems or project constraints that are not evident in the other materials already gathered.

Armed with all the information gathered in the design investigation, it is now time to start the design.

5.3 SYSTEM LAYOUT

The first step in storm sewer design is laying out the system. On the site topographic map, find the best location for the system to discharge to a stream or other body of water. You may need more than one discharge point. Next, locate the positions of the inlets which are placed at low points where they can most efficiently gather runoff. Finally, connect the inlets with pipes and check to confirm that their horizontal and vertical alignments do not interfere with other utilities or site components.

Selection of the discharge point is very important and not always easy. The most important principle guiding this decision is that the stormwater must be transferred from the pipe to the receiving water safely and without causing erosion damage or flooding. So, for instance, a discharge would not be placed in the middle of a field because the concentrated flow leaving the pipe would almost certainly erode the field. If a suitable stream is not available on the project site, you may be forced to extend your system a considerable distance off the site to find a proper discharge point. In Section 5.5, various methods of erosion prevention are discussed.

In many projects, an adjacent or nearby existing storm sewer may be utilized as the receiving body of water, thus eliminating the need to discharge to a stream. In such cases the proposed storm sewer connects to one of the existing inlets and, in effect, becomes part of the existing system. It may be necessary, however, to check the capacities of the existing pipes to verify that they can accept the additional discharge from the proposed system. Proper selection of the inlet locations is crucial to a good storm sewer design. In addition to choosing low points, many other factors must be considered. The most important of these are summarized below.

1. Grading—Obviously, the inlets must be placed where runoff can enter. This is accomplished by coordinating site grading with inlet positioning, that is, by shaping the ground to channel the runoff by gravity into the inlet. The most common way to accomplish this is by use of the curb or gutter along the edge of a road or parking area. Where no curb or gutter is available, such as in the middle of a lawn, the ground is graded into a swale.
2. Spacing—Inlets should be placed close enough together to prevent any single inlet from receiving too much runoff. A typical inlet can accept about 4 cfs (0.1 m³/s). When inlets are placed along a curbline, they should not be spaced farther apart than about 250 linear feet (75 linear meters) to prevent excessive flow along the gutter. This spacing also allows access to the pipes for inspection and maintenance. Conversely, inlets should not be placed too close together because such spacing is needlessly expensive.
3. Change of direction—Because storm sewer pipes must be straight, they must make sharp angles at every change of direction. In order to prevent clogging at the bend points and also to provide access to the pipes, special structures are provided at every change of direction. These structures typically are inlets or manholes.
4. Change of pipe slope or size—For the same reasons as change of direction, inlets or manholes must be placed at each point where the slope or size of the pipe changes.

Figure 5-5 Typical storm sewer layout for a street system.

Figure 5-6 Typical storm sewer layout for a site parking area.

Figure 5-5 and 5-6 show two examples of storm sewer system layout. In Figure 5-5, several principles of street layout are illustrated. First, runoff is channeled to the inlets by flowing along the gutters where it is intercepted by the inlets. Inlet 2 intercepts runoff from the land west of Road A as well as runoff from the westerly half of the pavement of Road A. Inlet 1, however, intercepts runoff from the easterly half of Road A and nowhere else. Even though Inlet 1 intercepts very little runoff compared to Inlet 2, it is included in the system to reduce the amount of gutter flow along the easterly side of Road A. Inlets 3 and 4 are placed at the uphill side of the intersection to prevent gutter flow from continuing across the intersection where it would interfere with traffic movement. Manhole 5 is included to provide a junction for two branches coming together. The pipe segments crossing the roads, that is, between Inlets 1 and 2 and Inlets 6 and 7 are called cross drains and serve the purpose of conveying the gutter flow from one side of the road to the side where the main sewer line is running.

In Figure 5-6, inlets again are arranged to intercept runoff flowing along the curb although in this case, the curb is part of the parking area for a building. Inlets must be placed at every low point in order to prevent puddling and they must be numerous enough to prevent excessive build-up of runoff flow in the parking area. Notice the pipes leading from the two corners of the building conveying runoff from the roof. These pipes could also convey water from a footing drain if one had been installed.

5.4 HYDRAULIC DESIGN

After laying out the system in plan view, you can begin the hydraulic design. The principle goal of hydraulic design is to determine the pipe size, segment by segment, throughout the system. To accomplish this, peak discharge, Q_p, must be computed for each pipe segment using the Rational Method.

Figure 5-7 shows the same storm sewer system depicted in Figure 5-5 but with delineated drainage areas tributary to the individual inlets. These areas must be delineated using topographic and site reconnaissance information. Also, hydraulic paths must be traced for each area in order to compute times of concentration.

Let's consider first the pipe segment from Inlet 1 to Inlet 2 which is the farthest upstream segment of the system. We will refer to this as Pipe Segment 1-2. All of the water flowing in this segment originates as runoff in its drainage area referred to as Area 1. Computation of Q_p is done using the procedure outlined in Section 4.5, including time of concentration, composite c value, etc. When Q_p is known, a size and slope is chosen for the pipe segment and its corresponding capacity determined. If the capacity is greater

Figure 5-7   Delineated drainage areas tributary to individual inlets of a typical storm sewer system.

than Q_p, the pipe segment is considered adequate and is accepted in the design. If the capacity is less than Q_p, the pipe segment is not adequate and a larger pipe size or steeper slope must be chosen.

In addition to checking pipe capacity, you must check inlet grate capacity. If Q_p exceeds 4 cfs (0.1 m³/s) you should either add another inlet at that location or shorten the distance between inlets thus reducing the drainage area. In the case of Inlet 1, Q_p will be much less than 4 cfs (0.1 m³/s) since its tributary drainage basin consists only of a portion of road surface.

Because the drainage Area 1 is so small, you will find that time of concentration is also small. The remotest point in the basin is located in the center of the road at the high point and most of the hydraulic path runs along the gutter. For very small basins like this, designers generally use a minimum value of time of concentration of 6 minutes. (Some designers use a higher value such as 10 minutes.) So, if a strict computation of t_c yields a value like 2.8 minutes, use 6.0 minutes in your Rational Method calculations. This does not reduce the validity of the overall design.

Next, let's consider the pipe segment from Inlet 2 to Inlet 3. Water flowing in this pipe originates from two separate sources: some comes from Pipe Segment 1-2 while the remainder enters directly into Inlet 2. To compute Q_p for this pipe, you must use a tributary drainage area equal to the sum of the areas for Inlet 1 and Inlet 2.

Figure 5-8 shows how drainage areas are cumulative as you work your way down the system. Figure 5-8(b) shows the drainage area for Pipe Segment 2-3. To compute time of concentration for this area, follow the procedure outlined in Section 4.3 in which the longest hydraulic path is chosen. Figure 5-9 shows two possible hydraulic paths for calculation of t_c for Pipe Segment 2-3. Notice that both points terminate at Inlet 2, which is the point of analysis for Pipe Segment 2-3. Since Path B includes overland flow in addition to gutter flow, it will undoubtedly be longer in time than Path A. Therefore, Path B is chosen.

Next, we turn our attention to Pipe Segment 3-5. Figure 5-8(c) shows the drainage area tributary to this pipe which includes the previous area plus the incremental area tributary to Inlet 3. To compute time of concentration for this area, choose the longest hydraulic path of the three possible paths shown in Figure 5-10. Notice that all three paths terminate at Inlet 3, which is the point of analysis for Pipe Segment 3-5. Either Path B or Path C could have a longer travel time. However, this can be determined only by computing t_c for both paths and comparing values. Whichever path yields the higher value of t_c will be chosen and that t_c will be used in calculating Q_p for Pipe Segment 3-5.

The next pipe segment is 4-5, but in this case we do not add the previous drainage areas as we did with segment 3-4 above. Pipe segment 4-5 is a branch of the overall storm sewer system and therefore does not carry the discharge from the upstream areas. In Figure 5-8(d), we see that the drainage area for this segment is confined to only the incremental area tributary to Inlet 4 referred to as Area 4. Time of concentration for this segment is simply the time of concentration for Area 4.

The next pipe segment is 5-7. Figure 5-8(e) shows that the drainage area tributary to this pipe includes the summation of all upstream incremental areas. No runoff is directly tributary to Pipe Segment 5-7 since Manhole 5 has no grate through which water can enter. So all discharge flowing in Pipe Segment 5-7 comes from the two upstream branches of the system. Manhole 5 represents the confluence of the two branches as well as the point of analysis for calculating Q_p for Pipe Segment 5-7. Time of concentration for this calculation is based upon the longest hydraulic path of the four possible paths originating in the four upstream incremental drainage areas.

The next pipe segment considered in the design is Segment 6-7. This is another branch segment (like Segment 4-5) and has a tributary drainage area as shown in Figure 5-8(f). Calculation of Q_p for this segment is similar to that for Segment 1-2, which has a comparable drainage area and time of concentration.

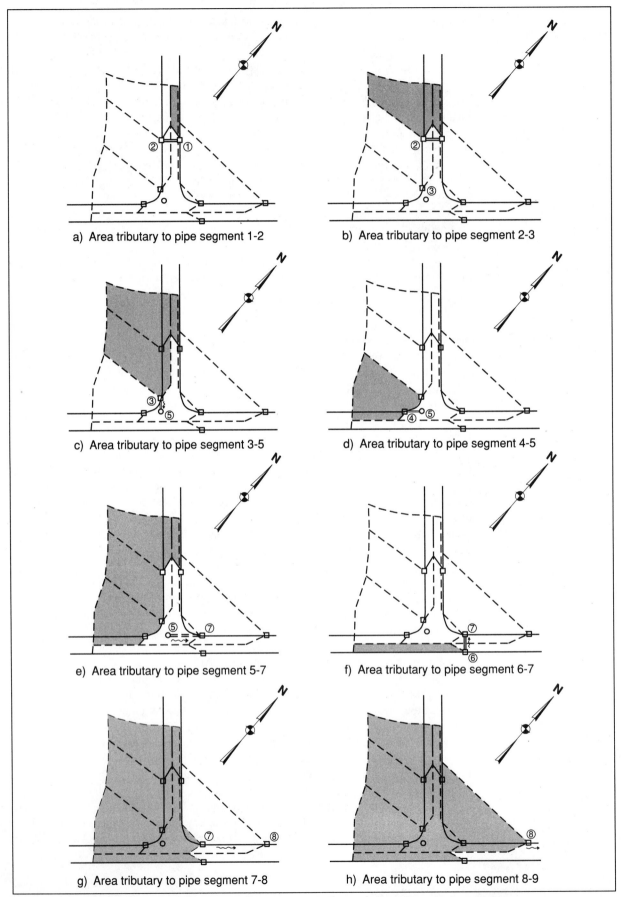

Figure 5-8   Drainage areas tributary to pipe segments through the storm sewer system.

Figure 5-9   Two possible hydraulic paths for calculating Q_p for pipe segment 2-3.

Figure 5-10   Three possible hydraulic paths for calculating Q_p for pipe segment 3-5.

The next pipe segment is 7-8 which carries water from three different sources:

1. Segment 6-7,
2. accumulated flow in Segment 5-7 and
3. runoff entering Inlet 7.

Figure 5-8 shows the tributary drainage area for this segment. Time of concentration is based upon the longest hydraulic path, terminating at Inlet 7, of the six possible paths originating in the six upstream incremental drainage areas.

The last pipe segment in the system we are analyzing is Segment 8-9. Figure 5-7 shows Inlet 8 and a pipe extending eastward toward Inlet 9, although Inlet 9 is not actually shown. Figure 5-8(h) shows the tributary drainage area for this segment which consists of the summation of all upstream incremental areas. Time of concentration for this segment is calculated in the same manner as the preceding segments, by selecting the hydraulic path which produces the longest travel time.

Example 5-1

Problem
Calculate the peak discharge in each pipe segment of the storm sewer system depicted in Figure 5-7 for a design storm frequency of 25 years. Assume that the project site is located in Pennsylvania, Region 1 and has the following parameters:

Inlet	Incremental Drainage Area (acres)	t_c* (min)	c**
1	0.07	6	.95
2	0.46	10	.45
3	0.52	10	.48
4	0.65	9	.41
5	—	—	—
6	0.10	6	.95
7	0.15	6	.95
8	0.70	14	.38

* time of concentration for each incremental drainage area
** composite runoff coefficient for each incremental drainage area

Solution

Pipe Segment 1-2
Since area and runoff coefficient are already known, find rainfall intensity by use of the I-D-F curves for Pennsylvania (Region 1) in Appendix C-3. Entering a duration of 6 minutes in the I-D-F graph, rainfall intensity for a 25-year storm is found to be

i = 5.5 in/h.

Peak runoff is then computed using Equation 4-2:

Q_p = Aci
Q_p = (.07) (.95) (5.5)
Q_p = 0.37 cfs.   (Answer)

Pipe Segment 2-3
Drainage area is equal to the sum of the upstream incremental areas:

$$A = 0.07 + 0.46$$
$$A = 0.53 \text{ acres.}$$

Composite runoff coefficient is the weighted average of the runoff coefficients of the upstream incremental areas:

$$c = \frac{(.07)(.95) + (.46)(.45)}{.53}$$

$$c = 0.52.$$

Time of concentration is based upon the hydraulic path with the longest travel time. Total time to Inlet 2 along hydraulic path A as shown in Figure 5-10 is 6.0 minutes plus the travel time in Pipe Segment 1-2. For the purpose of this example, assume this travel time is about 0.1 minute. Therefore, total time is 6.1 minutes. However, total time to Inlet 2 along hydraulic path B as shown in Figure 5-10 is 10 minutes (given). Thus, time of concentration is taken as

$$t_c = 10 \text{ min.}$$

Rainfall intensity is found in Appendix C-3 by entering a duration of 10 minutes. Therefore,

$$i = 4.3 \text{ in/h.}$$

Peak runoff is then computed using Equation 4-2:

$$Q_p = Aci$$
$$Q_p = (.53)(.52)(4.3)$$
$$Q_p = 1.2 \text{ cfs.} \qquad \text{(Answer)}$$

Pipe Segment 3-5
Drainage area is equal to the sum of the upstream incremental areas:

$$A = 0.07 + 0.46 + 0.52$$
$$A = 1.05 \text{ acres.}$$

Composite runoff coefficient is the weighted average of the runoff coefficients of the upstream incremental areas:

$$c = \frac{(.07)(.95) + (.46)(.45) + (.52)(.48)}{1.05}$$

$$c = 0.50.$$

Time of concentration is based upon the hydraulic path with the longest travel time. Hydraulic Path A was eliminated in the analysis for Segment 2-3. Total time along Path B to Inlet 3 is 10 minutes plus the travel time in Pipe Segment 2-3 which we will assume for now to be 0.5 minutes. Therefore, total time is 10.5 minutes. Total time to Inlet 3 along Hydraulic Path C as shown in Figure 5-10 is 10 minutes (given). Thus, time of concentration is taken as

$t_c = 10.5$ min.

Rainfall intensity is found in Appendix C-3 by entering a duration of 10.5 minutes. Therefore,

   $i = 4.2$ in/h.

Peak runoff is then computed using Equation 4-2:

   $Q_p = Aci$
   $Q_p = (1.05)\,(.50)\,(4.2)$
   $Q_p = 2.2$ cfs.   (Answer)

Pipe Segment 4-5
This segment is a branch of the system and, as shown earlier, is treated as though it is the beginning of the system. Area and runoff coefficient are given. Rainfall intensity is found in Appendix C-3 by entering a duration of 9 minutes. Therefore,

   $i = 4.5$ in/h.

Peak runoff is then computed using Equation 4-2:

   $Q_p = Aci$
   $Q_p = (.65)\,(.41)\,(4.5)$
   $Q_p = 1.2$ cfs.   (Answer)

Pipe Segment 5-7
Drainage area is equal to the sum of the upstream incremental areas:

   $A = .07 + .46 + .52 + .65$
   $A = 1.70$ acres.

Composite runoff coefficient is the weighted average of the runoff coefficients of the upstream incremental areas:

$$c = \frac{(.07)\,(.95) + (.46)\,(.45) + (.52)\,(.48) + (.65)\,(.41)}{1.70}$$

   $c = 0.46$.

Time of concentration is based upon Hydraulic Path B because it requires the greatest amount of time. We saw earlier that this path has a time of 10.5 minutes from its beginning to Inlet 3. However, for Pipe Segment 5-7, the point of analysis is Manhole 5. So, the time of concentration is 10.5 minutes plus the travel time in Pipe Segment 3-5, which we will show later to be 0.1 minute. Therefore, total time is 10.6 minutes. Total time to Manhole 5 along any other path is less than 10.6 minutes. Thus, time of concentration is taken as

   $t_c = 10.6$ minutes.

Rainfall intensity is found in Appendix C-3 by entering a duration of 10.6 minutes. Therefore,

$i = 4.2$ in/h.

Peak runoff is then computed using Equation 4-2:

$Q_p = Aci$
$Q_p = (1.70) (.46) (4.2)$
$Q_p = 3.3$ cfs.   (Answer)

Pipe Segment 6-7
This segment is another branch of the system and therefore treated as though it is at the beginning. Area and runoff coefficients are given. Rainfall intensity for a duration of 6 minutes was already found to be

$i = 5.5$ in/h.

Peak runoff is then computed using Equation 4-2:

$Q_p = Aci$
$Q_p = (.10) (.95) (5.5)$
$Q_p = 0.52$ cfs.   (Answer)

Pipe Segment 7-8
Drainage area is equal to the sum of the upstream incremental areas:

$A = 1.70 + .10 + .15$
$A = 1.95$ acres.

Notice that in this computation, the first four incremental areas are summed in the first term: 1.70 acres. Composite runoff coefficient is a weighted average as computed previously:

$$c = \frac{(1.70)(.46) + (.10)(.95) + (.15)(.95)}{1.95}$$

$c = 0.52.$

Notice that in this computation, the first four incremental areas were taken as one composite c value. This gives the same result as listing them separately. Time of concentration is once again based upon Hydraulic Path B which has a total time to Inlet 7 of 10.6 minutes plus the travel time in Pipe Segment 5-7 which we will show later to be 0.2 minutes. Therefore, total time is 10.8 minutes. Total time to Inlet 7 along any other path is less than 10.8 minutes. Thus, time of concentration is taken as

$t_c = 10.8$ min.

Rainfall intensity is found in Appendix C-3 by entering a duration of 10.8 minutes. Therefore,

$i = 4.1$ in/h.

Peak runoff is then computed using Equation 4-2:

$Q_p = Aci$
$Q_p = (1.95)\,(.52)\,(4.1)$
$Q_p = 4.2$ cfs.   (Answer)

Pipe Segment 8-9
Drainage area is equal to the sum of the upstream incremental areas:

$A = 1.95 + 0.70$
$A = 2.65$ acres.

Composite runoff coefficient is a weighted average as computed previously:

$$c = \frac{(1.95)\,(.52) + (.70)\,(.38)}{2.65}$$

$c = 0.48.$

Time of concentration is, as always, based upon the longest hydraulic path. Using Inlet 8 as point of analysis, Path B has a time of 10.8 minutes plus travel time in Pipe Segment 7-8 which we will show later to be 0.9 minutes. Therefore, total time for Path B is 11.7 minutes. However, total time for the hydraulic path within the incremental area tributary to Inlet 8 is 14 minutes. So this is now the longest hydraulic path to the point of analysis and therefore time of concentration is taken as

$t_c = 14$ min.

Rainfall intensity is found in Appendix C-3 by entering a duration of 14 minutes. Therefore,

$i = 3.75$ in/h.

Peak runoff is then computed using Equation 4-2:

$Q_p = Aci$
$Q_p = (2.65)\,(.48)\,(3.75)$
$Q_p = 4.8$ cfs.   (Answer)

In performing these design calculations, you should be aware of peak runoff entering each individual inlet. In this case, it is evident that no single incremental area produces a peak runoff greater than 4 cfs. If we had suspected one area of exceeding its inlet capacity, we would have checked peak runoff for that area.

Peak runoff for Pipe Segment 8-9 represents the total peak runoff for the storm sewer system to that point (Inlet 8). If we had computed peak runoff for each incremental area and then added them together, the resulting total peak runoff would be greater than the 4.8 cfs computed. It would be incorrect to add the individual peak runoff values because such a procedure does not account for attenuation as the runoff travels from each incremental area to Inlet 8.

Travel times in pipe segments used in Example 5-1 are based upon computations which will be explained in Example 5-2. The calculations carried out in Example 5-1 can be conveniently arranged in tabular form as shown in Table 5-1.

Table 5-1
Runoff Calculations for the Pipe Segments in Example 5-1.

(1) Pipe Segment	(2) Incr. Area (acres)	(3) c	(4) Incr. Ac (acres)	(5) Cumm. Ac (acres)	(6) t_c (min.)	(7) i (in/h)	(8) Q_p (cfs)
1-2	0.07	.95	.067	.067	6	5.5	0.37
2-3	0.46	.45	.207	.274	10	4.3	1.2
3-5	0.52	.48	.250	.524	10.5	4.25	2.2
4-5	0.65	.41	.267	.267	9	4.5	1.2
5-7	—	—	—	.791	10.6	4.25	3.3
6-7	0.10	.95	.095	.095	6	5.5	0.52
7-8	0.15	.95	.143	1.029	10.8	4.2	4.2
8-9	0.70	.38	.266	1.295	14	3.75	4.8

Look this table over and notice column 4 labeled *Incremental Ac*. Each number in this column is the product of the corresponding numbers in columns 2 and 3. Column 5 then shows cumulative values from column 4 which accomplishes two computations at once: it computes cumulative area and composite c for the accumulated area. These operations were done separately earlier in the example.

The desired result, peak flow for each pipe segment, is shown in column 8. Each value in column 8 is computed by multiplying the corresponding values in columns 5 and 7. This multiplication is equivalent to using Equation 4-2 as we did throughout the example.

In order to better understand how the computations were done, let's analyze column 5 a little closer. Each value represents the total area tributary to the particular pipe segment multiplied by the composite runoff coefficient of that area. The first three values are running totals of the incremental values in column 4. The fourth value, however, is not a running total because Pipe Segment 4-5 is a branch of the system and as such carries runoff solely from its incremental drainage area.

Now, let's take a similar look at column 6. Notice that Pipe Segments 2-3, 3-5, 5-7 and 7-8 form a sequence of t_c's in which each t_c increases by an amount equal to the travel time in the previous pipe segment. Pipe Segments 4-5 and 6-7 are not in this sequence because they are branches and are not on the main stem of the system. Segment 8-9 is also not in the sequence despite being part of the main stem simply because a different hydraulic path was used to calculate time of concentration.

Each value in column 7 is simply a rainfall intensity found in Appendix C-3 corresponding to the time of concentration in column 6.

Charts of this type typically are utilized as a format for storm sewer design calculations and can be prepared using spreadsheet software.

Pipe Size Computation

After computing peak discharge flowing in each pipe segment, the next step in storm sewer design is to choose an appropriate size for the pipe. The slope of the pipe is chosen at this time as well since both size and slope are needed to determine capacity. Also important to capacity is the type of pipe: for example, concrete, corrugated metal or plastic. Usually the pipe's material is selected first since this decision is based primarily upon cost and other factors.

Pipe slope is dictated to a great extent by surface grades although some variation is available to the designer. So at this point in the design, a profile of the proposed sewer is very helpful if not essential. Plotting a profile reveals not only the slope but also the pipe depth and any potential conflicts with other utilities.

Figure 5-11 Profile of the storm sewer system depicted in Figure 5-7.

Figure 5-11 shows a profile of the storm sewer system shown in Figure 5-7 and which is the subject of Examples 5-1 and 5-2. Notice that the profile is drawn with an exaggerated vertical scale. That is, vertical distances are ten times larger than horizontal distances. Profiles typically are drawn in this fashion to render very subtle slopes more obvious and easier to employ in design.

In performing the design, a profile like that in Figure 5-11 is drawn on a work sheet and the pipe segments added one at a time as their sizes are determined by the computations. It is a trial and error process in which a 12-inch pipe is chosen first and its capacity compared to Q_p for that segment. If necessary, a larger size is then chosen. The following example illustrates the design process.

Example 5-2

Problem
Determine the pipe sizes for the storm sewer system depicted in Figure 5-7 for a design storm frequency of 25 years. Assume that the pipes are to be reinforced concrete pipes (RCP) with Manning's n-value of 0.015. Also, assume that the system has the following parameters:

Pipe Segment	Length (ft)	Slope (%)
1-2	30	2.0
2-3	200	3.25
3-5	25	2.5
4-5	25	2.0
5-7	50	0.5
6-7	30	2.0
7-8	220	0.5
8-9	—	0.5

Solution

Pipe Segment 1-2
Start by assuming a size of 12 inches which we will take as the minimum size for this example. Now determine the capacity for a 12-inch diameter pipe with n = 0.015 and slope = 2.0% using Pipeflow Chart 35 in Appendix A-4. Locate the slope line (0.02 ft/ft) and find the upper end. From that point, trace straight down to the *Discharge* axis for n = 0.015 and read the answer: Q = 4.4 cfs.

Since this capacity is greater than the computed peak runoff of 0.37 cfs, the assumed pipe size is accepted.

Pipe Segment 2-3
Find the capacity for a 12-inch pipe at a slope of 3.25% using Chart 35. Interpolate between the slope lines for 0.03 ft/ft and 0.04 ft/ft and trace down to the discharge axis and read the answer: Q = 5.8 cfs.

Since this capacity is greater than the computed peak runoff of 1.2 cfs, the assumed pipe size is accepted.

Pipe Segment 3-5
Find the capacity for a 12-inch pipe at a slope of 2.5% using Chart 35 Appendix A-4. Interpolate between the slope lines for 0.02 ft/ft and 0.03 ft/ft and read the answer: Q = 5.0 cfs.

Since this capacity is greater than the computed peak runoff of 2.2 cfs, the assumed pipe size is accepted.

Pipe Segment 4-5
This segment uses the same pipe size and slope as Segment 1-2 so it has the same capacity, namely 4.4 cfs.

Since this capacity is greater than the computed peak runoff of 1.2 cfs, the assumed 12-inch pipe size is accepted.

Pipe Segment 5-7

Again, start by assuming a pipe size of 12 inches. The capacity for a 12-inch pipe with slope of 0.5% is found in Chart 35 to be Q = 2.2 cfs. Since this capacity is less than the computed peak runoff of 3.3 cfs, the assumed pipe size is not accepted.

Next, choose a pipe size of 15 inches. The capacity for a 15-inch pipe with slope of 0.5% is found in Chart 36 to be Q=4.0 cfs. Since this capacity is greater than the computed peak runoff of 3.3 cfs, the 15-inch pipe is accepted.

Pipe Segment 6-7

This segment uses the same pipe size and slope as Segments 1-2 and 4-5 so it has the same capacity of 4.4 cfs. This capacity is greater than the computed peak runoff of 0.52 cfs and so a 12-inch pipe is selected.

Pipe Segment 7-8

Start by assuming a pipe size of 15 inches since pipe sizes must be equal to or greater than upstream pipes. The capacity for a 15-inch pipe with a 0.5% slope was determined previously to be 4.0 cfs.

Since this capacity is less than the computed peak runoff of 4.2 cfs, the assumed 15-inch pipe is not accepted and an 18-inch pipe is tried next. The capacity for an 18-inch pipe with a slope of 0.5% is found in Chart 37 to be Q = 6.4 cfs.

Since 6.4 cfs is greater than 4.2 cfs, the 18-inch pipe is accepted.

Pipe Segment 8-9

Start by assuming a pipe size of 18 inches since pipe sizes must be equal to or greater than upstream pipes. The capacity of an 18-inch pipe with a slope of 0.5% was already determined to be 6.4 cfs.

Since 6.4 cfs is greater than 4.8 cfs, the 18-inch pipe is accepted. This hydraulic design can be presented in chart form as shown in Figure 5-12.

A close and detailed scrutiny of Figure 5-12 is important in understanding the storm sewer design process. This chart is an exact facsimile of those used universally by engineers for presentation of storm sewer design. Notice first that columns 1 through 8 are the same as the chart shown in Example 5-1. So, let's now focus on columns 9 through 14.

These columns are arranged in the order in which the numbers are determined. First, pipe length is taken from the layout plan. Next, slope is determined by drawing a trial pipe on the profile worksheet. The size is also determined in the trial. Now that size and slope have been postulated, capacity is determined from the pipe flow charts and the proposed size is accepted.

The remaining two columns are included to allow a computation of travel time in the pipe segment which is needed to determine subsequent time of concentration. This is the computation referred to in Example 5-1.

Look first at column 13, Segment 1-2. The velocity of 3.4 ft/s was found in Chart 35 in Appendix A-4. This is the average velocity in Segment 1-2 for a discharge of 0.37 cfs. It is determined by entering the chart at a discharge of 0.37 cfs and tracing upward to the slope line of 0.02 ft/ft. From that point, trace to the left (keeping parallel to the velocity lines) to the velocity scale and interpolate between 3 ft/s and 4 ft/s.

Travel time in Pipe Segment 1-2 is then computed using the formula

$$\text{time} = \frac{\text{distance}}{\text{velocity}}$$

DRAINAGE CALCULATIONS
25-year storm
n = 0.015

(1) Pipe Segment From	(1) Pipe Segment To	(2) A—Increm. Area (acres)	(3) C—Runoff Coefficient	(4) A·c—Increm.	(5) A·c—Cumul.	(6) t_c—Time of Conc. (min.)	(7) i—Rainfall Intensity (in/h)	(8) Q_p—Peak Runoff (cfs)	(9) Pipe Length (ft)	(10) Slope (%)	(11) Size (in)	(12) Capacity (full) (cfs)	(13) Velocity (fps) (Design Flow)	(14) Travel Time in Pipe (min)
1	2	0.07	.95	.067	.067	6	5.5	0.37	30	2.0	12	4.4	3.4	0.15
2	3	0.46	.45	.207	.274	10	4.3	1.2	200	3.25	12	5.8	5.6	0.6
3	5	0.52	.48	.250	.524	10.6	4.25	2.2	25	2.5	12	5.0	6.0	0.1
4	5	0.65	.41	.267	.267	9	4.5	1.2	25	2.0	12	4.4	4.8	0.1
5	7	—	—	—	.791	10.7	4.25	3.3	50	0.50	15	4.0	3.6	0.2
6	7	0.10	.95	.095	.095	6	5.5	0.52	30	2.0	12	4.4	3.7	0.1
7	8	0.15	.95	.143	1.029	10.9	4.2	4.2	220	0.50	18	6.4	3.9	0.9
8	9	0.70	.38	.266	1.295	14	3.75	4.8	—	0.50	18	6.4	4.0	—

Figure 5-12 Drainage calculation chart for Example 5-2.

where distance, in this case, is the pipe length of 30 ft. Therefore,

$$\text{time} = \frac{30}{3.4}$$

time = 8.82 seconds

time = 0.15 minutes.

The travel time is then entered in column 14. Each number in column 14 represents the travel time of the discharge in the corresponding pipe segment. These travel times are used to compute time of concentration as the design progresses down the system.

For instance, the time of concentration in column 6 for Segment 3-5 is 10.6 min, which is computed by adding 10 min (column 6, segment 2-3) to 0.6 min (column 14, segment 2-3).

As another example, the time of concentration shown in column 6 for Segment 5-7 is 10.7 min, which is computed by adding 10.6 min (column 6, segment 3-5) to 0.1 min (column 14, segment 3-5).

Of course, not all t_c values in column 6 are computed in this manner. Exceptions include segments with no upstream tributary pipes (Segments 1-2, 4-5, 6-7) and segments with hydraulic paths originating in their own incremental drainage area (Segments 2-3 and 8-9).

5.5 STORM SEWER OUTFALLS

Careful attention must be given to the design of the storm sewer outfall or point of discharge from the system. This is the point where collected stormwater is discharged from the system to the receiving body of water, and it is here that most soil erosion damage can occur.

One of the first considerations in outfall design is the outfall structure. A storm sewer pipe can simply emerge from the ground and terminate with no structure, or a headwall can be constructed at the discharge point. Figure 5-13 depicts in isometric views several outfall structures which are in general use today. In Figure 5-14, the structures are shown as they might appear on construction plans.

The purpose of a headwall or other outfall structure is to protect the soil around the discharge pipe from erosion and to keep the slope in place. The outfall structure actually acts as a small retaining wall. Very few cases occur when a structure can be completely avoided.

In some cases, materials other than those shown in Figure 5-13 can be used to construct a headwall. Examples include gabions, grouted stone and railroad ties.

Generally, storm sewer outfalls are configured in one of three ways as described below and illustrated in Figure 5-15.

1. *Outfall at stream bank.* In this case, the storm sewer pipe extends all the way to the receiving stream and the headwall is constructed in the bank.

2. *Channel connecting outfall with stream.* In this case, the headwall or flared end section is located several feet from the stream and a channel is constructed to connect the two.

3. *Outfall discharging onto stream overbank.* This arrangement is similar to item 2 above except no channel is cut between the outfall and the stream. Discharge is allowed to flow overland across the overbank and eventually into the stream. This arrangement is selected often when the overbank area is environmentally sensitive as in the case of *wetlands* and excavating a channel is undesirable.

a) Concrete headwall with wingwalls

b) Concrete straight headwall

c) Flared end section (pre-formed flared shape
connected to end of discharge pipe; materials
include concrete, corrugated metal and plastic.)

Figure 5-13   Examples of typical outfall structures.

PLAN

OUTLET END

a) Concrete headwall and apron.

b) Concrete flared end section.

Figure 5-14 Construction details of various outfall structures.

a) Outfall at stream bank

b) Channel connecting outfall with stream

c) Outfall discharging onto stream overbank

Figure 5-15   Schematic representation of three types of storm sewer outfalls.

Figure 5-16 Storm sewer outfall with severely eroded ground immediately downstream.

Discharge exiting from the outfall is in concentrated form and therefore potentially damaging to unprotected ground. Excessive velocity can damage the ground by dislodging soil particles and washing them away. Figure 5-16 shows an outfall with severely eroded ground immediately downstream.

An important part of the design process is the selection of a method for protecting the ground surface wherever velocities are destructive, such as at the outfall. Appendix A-2 shows a list of maximum allowable velocities for various types of ground cover. These velocities are a guide to use when designing an outfall.

Several methods are available for the control of erosion at a storm sewer outfall including reduction of discharge velocity and covering the ground with an erosion resistant lining. These methods are summarized as follows.

1. *Reduce discharge velocity.* This can be accomplished simply by reducing the slope of the last pipe segment before the outfall.

2. *Energy dissipator.* In cases of very high velocity, specially designed obstruction blocks can be placed at the outlet to create a head loss and therefore reduction of velocity. Such a design is beyond the scope of this book.

3. *Stilling basin.* A depression in the ground surface can be provided at the outlet to absorb excessive energy of the discharge. Such a basin is designed to trap water and therefore must have a provision for drainage between storms. This design is also beyond the scope of this book.

4. *Riprap.* This is a lining of heavy rocks covering the vulnerable ground to both protect the ground surface and slow the discharging velocity at the same time. Riprap design procedures can be found in many publications, especially agency design manuals.

5. *Erosion control mats.* A wide variety of commercially available products consist of mesh netting which is placed on the ground to protect the surface and to anchor a vegetative cover as it grows.

6. *Sod.* This is simply the application of sod strips to the vulnerable ground to provide a grass lining without the disadvantage of growing grass from seed. This protective lining is generally effective but expensive and labor intensive.

7. *Gabions.* Gabions are rectangular wire mesh baskets filled with rocks and placed on the ground as a protective lining similar to riprap.

Riprap

Riprap is a common erosion control lining used at storm sewer outfalls as well as culvert outlets, especially in areas where suitable rock materials are readily available. Figure 5-17 depicts a storm sewer outfall with riprap protection in place. A lining of the ground at a pipe outlet is generally referred to as an *apron*.

Design of riprap outfall protection includes many factors, the most basic of which are listed below:

1. Type of stone
2. Size of stone
3. Thickness of stone lining
4. Length of apron
5. Width of apron

Procedures for determining these parameters are contained in many published design manuals in use today across the United States. When designing riprap, you should be aware of any design standard used by the agency reviewing your design and consider employing same. However, a simple riprap design procedure is presented here for general use.

1. *Type of stone.* The stones used for riprap should be hard, durable and angular. Angularity, which is a feature of crushed stone from a quarry, helps keep the stones locked together when subjected to the force of moving water.

Figure 5-17   Storm sewer outfall with riprap erosion protection in place.

2. *Size of stone.* The stones should be well-graded in a range of sizes referred to as *gradation.* Gradation is another factor promoting the interlocking of the stones. Size is defined as the median diameter, d_{50}, which is the diameter of stones of which 50% are finer by weight, and is selected using the following formulas:

$$d_{50} = \frac{.02}{TW} \left(\frac{Q}{D_0} \right)^{4/3} \quad \text{(English Units)} \tag{5-1}$$

where d_{50} = median stone size, ft,
Q = design discharge, cfs,
D_0 = maximum pipe or culvert width, ft,
TW = tailwater depth, ft.

$$d_{50} = \frac{.044}{TW} \left(\frac{Q}{D_0} \right)^{4/3} \quad \text{(SI Units)} \tag{5-1a}$$

where d_{50} = median stone size, m,
Q = design discharge, m³/s,
D_0 = maximum pipe or culvert width, m,
TW = tailwater depth, m.

Note: For this riprap design, tailwater is the flow depth in the pipe at the outlet.

3. *Thickness of stone lining.* The blanket of stones should be three times the median stone size if no filter fabric liner between the stones and the ground is used. If a filter fabric liner is used, the thickness should be twice the median stone size.

4. *Length of apron.* The length, L_a, of the apron is computed using one of the formulas below.

If the design tailwater depth, TW, is greater than or equal to $\frac{1}{2}D_0$, then

$$L_a = \frac{3Q}{D_0^{3/2}} \quad \text{(English Units)} \tag{5-2}$$

where L_a = apron length, ft,
Q = design discharge, cfs,
D_0 = maximum pipe or culvert width, ft.

$$L_a = \frac{5.4Q}{D_0^{3/2}} \quad \text{(SI Units)} \tag{5-2a}$$

where L_a = apron length, m,
Q = design discharge, m³/s,
D_0 = maximum pipe or culvert width, m.

If TW (design) is less than $\frac{1}{2}D_0$, then

$$L_a = \frac{1.8Q}{D_0^{3/2}} + 7D_0 \quad \text{(English Units)} \tag{5-3}$$

where L_a = apron length, ft,
Q = design discharge, cfs,
D_0 = maximum pipe or culvert width, ft.

$$L_a = \frac{3.26Q}{D_0^{3/2}} + 7D_0 \quad \text{(SI Units)} \tag{5-3a}$$

where L_a = apron length, m,
 Q = design discharge, m³/s,
 D_0 = maximum pipe or culvert width, m.

Figure 5-18 illustrates the dimensions involved with apron size.

Figure 5-18   Dimensions of riprap apron of pipe or culvert outlet.

5. *Width of apron.* If a channel exists downstream of the outlet, the riprap width is dictated by the width of the channel. Riprap should line the bottom of the channel and part of the side slopes. The lining should extend one foot (0.3 m) above the design tailwater depth. The extra 1-foot height is called *freeboard* and is employed as a safety measure.

If no channel exists downstream of the outlet, the width, W, of the apron is computed using one of the two formulas below.

If the design tailwater depth, TW is greater than or equal to $\frac{1}{2}D_0$, then

$$W = 3D_0 + 0.4L_a \qquad (5\text{-}4)$$

where W = apron width, ft (m),
       D_0 = max. pipe or culvert width, ft (m),
       L_a = apron length, ft (m).

If TW (design) is less than $\frac{1}{2}D_0$, then

$$W = 3D_0 + L_a \qquad (5\text{-}5)$$

where W = apron width, ft (m),
       D_0 = max. pipe or culvert width, ft (m),
       L_a = apron length, ft (m).

Example 5-3

Problem

A storm sewer outfall is to be designed for direct discharge to the ground (no channel). Design discharge is 70 cfs and the last pipe segment is a 36-inch reinforced concrete pipe (RCP) with a 1.3% slope. The ground downstream of the outfall is predominantly silt loam.

Solution

First, determine the velocity of the design discharge at the outfall. Using Chart 43 of Appendix A-4, design velocity is found to be 10.5 fps. (The Manning's n-value is assumed to be 0.012.)

Next, compare the design velocity to the permissible velocity found in Appendix A-2. For silt loam, the permissible velocity is 3.0 fps.

Since design velocity exceeds permissible velocity, a protective lining will be required. Riprap is chosen based upon cost and accessibility of the stone.

To design the riprap apron, first determine the size, d_{50}, using Equation 5-1. Before using Equation 5-1, TW is determined from Chart 43 to be 2.2 feet. Then, using Equation 5-1,

$$d_{50} = \frac{.02}{2.2}\left(\frac{70}{3}\right)^{4/3}$$

$$d_{50} = 0.61 \text{ ft } (7\tfrac{1}{4} \text{ inches}).$$

Specify 8-inch stone since it is the next higher whole inch.

Next, determine the apron length, L_a. Since TW = 2.2 ft which is greater than $\frac{1}{2}D_0$, use Equation 5-2:

$$L_a = \frac{(30)\,(70)}{(3)^{3/2}}$$

$$L_a = 40.4 \text{ ft.}$$

Specify 40 feet since it is the rounded value of L_a.

Next, since there is no channel downstream of the outfall, determine the apron width, W. Since TW > $\frac{1}{2}D_0$, use Equation 5-4:

$$W = 3(3) + 0.4(40.4)$$

$$W = 25.2 \text{ ft.}$$

Specify 25 ft since it is the rounded value of W.

The outfall design is summarized in Figure 5-19.

Figure 5-19   Outfall design for Example 5-3.

5.6 CASE STUDY

In this case, a residential subdivision is proposed for a 28.5 acre tract in suburban Atlanta, Georgia. The subdivision, named Tall Pines, consists of 25 building lots plus one additional lot at the northerly end for a detention basin. As shown in Figure 5-20, one principle road, Road A, traverses the tract from south to north, and two stub roads, Road B and Road C, intersect Road A and run to opposite property lines.

In this case study, we will analyze the storm sewer design for the subdivision, and in another case study in Chapter 7 we will analyze the detention design.

To analyze the storm sewer design, we will trace the following procedures normally included in such design:

1. Design investigation
2. System Layout
3. Hydraulic design
4. Outfall design

The design investigation resulted in a topographic map used for the base map in Figure 5-20. Also, a site visit revealed that the tract is completely wooded in its existing condition. A meeting with the municipal engineer confirmed that the existing storm sewer in Tall Pines Road has sufficient capacity to accept additional discharge from the tract.

The subdivision was then designed in accordance with municipal zoning requirements. Roads A and B were proposed to extend to the property lines to provide access to adjoining tracts for future development.

Next the storm sewer system was laid out in plan view to locate key drainage structures. Figures 5-21 and 5-22 show plans and profiles of the roads, including inlet and pipe segment locations. Since Road A has a high point at about Station 6+55, drainage must flow in two opposite directions. Some runoff flows southerly and is picked up by inlets at Station 0 + 75 while the remainder flows northerly toward the existing stream. Since Road B has a curbed cul-de-sac with a low point, an inlet was placed at its end to pick up all runoff flowing down the street.

No inlet is placed at the end of Road C even though it is pitched toward the adjoining property to the west. Any runoff flowing down Road C and onto adjoining property is kept to a minimum by placing an inlet at Station 9 + 93 to intercept discharge before it turns the corner. Normally runoff in concentrated form should not be allowed to run on the ground onto another property because of the danger of erosion. But if the amount of flow can be reduced to a minor level, it may be allowed. Nonetheless, if the adjoining property owner objected, it would be necessary to design measures to eliminate the flow.

The pair of inlets at Station 9 + 93 are piped northerly down Road A to the cul-de-sac and then into the detention basin. The inlet at the end of Road B cannot be piped westerly toward Road A because Road B is pitched in the opposite direction, as shown in Figure 5-22. Therefore, the inlet is piped northerly to an outfall and then a drainage ditch or swale runs northerly along the easterly property line to the detention basin. Figure 5-23 shows a profile of the drainage swale.

Figure 5-20   Map of Tall Pines subdivision located near Atlanta, Georgia.
(Courtesy of Jaman Engineering.)

Figure 5-21 Plan and profile of Road A, Tall Pines subdivision. *(Courtesy of Jaman Engineering.)*

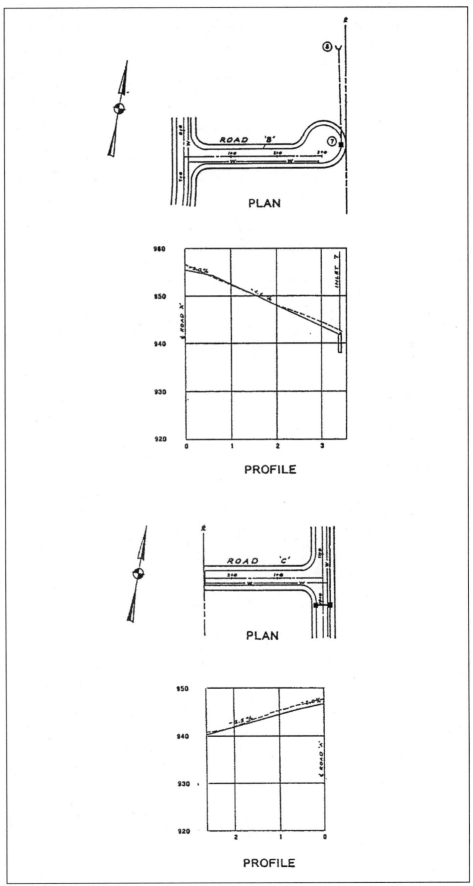

Figure 5-22   Plan and profile of Roads B and C, Tall Pines subdivision. *(Courtesy of Jaman Engineering.)*

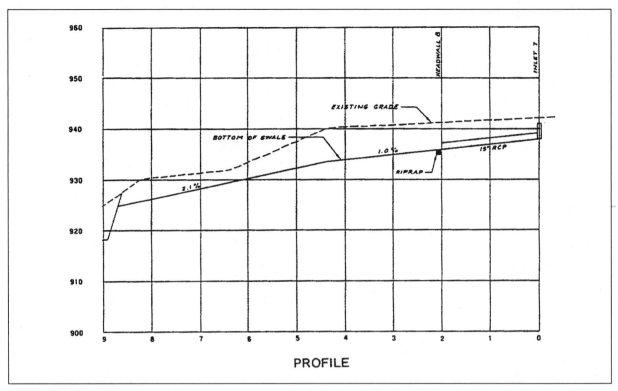

Figure 5-23 Profile of drainage swale and pipe segment 7-8. *(Courtesy of Jaman Engineering.)*

After completing the storm sewer layout, hydraulic design was initiated. Figure 5-24 shows the drainage area map for the subdivision, with incremental drainage basins delineated for each pipe segment. A close look at Figure 5-24 reveals that some of the basin divides do not follow the existing contours. This is because proposed grading will alter the contour lines and it is the proposed contours that dictate basin divides. For example, in some lots, basin divides run along the rear sides of proposed houses because the lots will be graded to pitch toward the road. In other lots, such as Lots 15 through 18, it was impractical to place enough fill to pitch the lawns to the road, so these lots drain away from the road.

To start the hydraulic design process, key parameters were selected based upon good design practice and local regulations. These include the following:

Design storm frequency:	25-year
Pipe material:	reinforced concrete
Pipe Manning n-value:	0.012
Min. pipe size	15 inches
Runoff coefficients:	impervious—0.90
	lawn—0.30
	woods—0.20

Next drainage areas were measured, times of concentration computed and a table of results prepared.

Figure 5-24 Drainage area map of Tall Pines subdivision. *(Courtesy of Jaman Engineering.)*

Times of concentration were determined by adding overland flow time and gutter flow time. Overland flow times were taken from Appendix C-2 using average grass surface, and gutter flow times were taken from Figure 4-10 using paved surface. Table 5-2 shows a summary of t_c computations and Table 5-3 presents a summary of incremental areas, composite c-values and times of concentration.

		Table 5-2 Computation of t_c for Tall Pines Subdivision	
Inlet		**Time of Concentration**	
1	Overland Gutter	220' @ 5.5% 150' @ 6.0%	$t_1 = 14$ min. $t_2 = 0.5$ min. $t_c = 14.5$ min.
2	Overland Gutter	90' @ 2.0% 260' @ 4.7%	$t_1 = 12$ min. $t_2 = 1.0$ min. $t_c = 13.0$ min.
3	Overland Gutter	110' @ 7.0% 250' @ 8.0%	$t_1 = 10.5$ min. $t_2 = 0.7$ min. $t_c = 12.2$ min.
4	Overland Gutter	250' @ 5.8% 200' @ 8.0%	$t_1 = 15$ min. $t_2 = 0.6$ min. $t_c = 15.6$ min.
5	Overland Gutter	290' @ 8.5% 110' @ 3.0%	$t_1 = 14.5$ min. $t_2 = 0.5$ min. $t_c = 16.0$ min.
7	Overland Gutter	250' @ 3.4% 80' @ 3.4%	$t_1 = 16.5$ min. $t_2 = 0.4$ min. $t_c = 16.9$ min.
9	Overland Gutter	230' @ 6.0% 280' @ 2.7%	$t_1 = 14$ min. $t_2 = 1.4$ min. $t_c = 15.4$ min.
10	Overland Gutter	100' @ 2.0% 130' @ 2.7%	$t_1 = 12.5$ min. $t_2 = 0.6$ min. $t_c = 13.1$ min.

Table 5-3
Summary of Runoff Parameters for Tall Pines Subdivision

Inlet or Manhole	Incremental Drainage Area (acres)	Impervious Area (acres)	Lawn Area (acres)	Wooded Area (acres)	Composite c	t_c (min.)
1	.98	.24	.55	.19	.43	14.5
2	.83	.24	.54	.05	.47	13.0
3	.74	.28	.46	—	.53	12.2
4	1.12	.22	.57	.33	.39	15.6
5	.64	.06	.14	.44	.29	16.0
6	—	—	—	—	—	—
7	1.70	.51	.79	.40	.46	16.9
8	—	—	—	—	—	—
9	1.08	.30	.62	.16	.45	15.4
10	.46	.14	.32	—	.48	13.1
11	.12	.07	.05	—	.65	*
12	.64	.21	.43	—	.50	*
13	.34	.10	.24	—	.48	*
14	.31	.17	.14	—	.63	*
15	.52	.18	.34	—	.51	*
	9.04	2.66	5.04	1.34		

*These times of concentration are, by inspection, less than any upstream time of concentration, and therefore not needed.

Storm sewers in this subdivision are divided into three separate systems, each corresponding to a different discharge or outfall point. Each system was designed separately starting with the southerly system and results were recorded in the design table shown in Figure 5-25.

Southerly System

Pipe Segment 1-2
Peak runoff tributary to this segment is computed in columns 2 through 8 in Figure 5-25. The drainage area tributary to Segment 1-2 is called Incremental Area 1. The value Ac shown in column 5 is the product of incremental area, A, and runoff coefficient, c. Time of concentration of 14.5 minutes shown in column 6 is that computed previously for Incremental Area 1. Rainfall intensity of 6.1 in/h shown in column 7 was determined from the rainfall intensity duration curves for Atlanta depicted in Appendix C-3. Peak runoff of 2.6 cfs shown in column 8 was computed by multiplying the values in columns 5 and 7.

Segment 1-2 is a cross drain and is assigned an arbitrary slope of 2.0% and minimum size of 15 inches. A 2.0% slope is used because it assures adequate capacity with sufficient vertical drop for ease of construction. Of course, the slope could be reduced if less drop is available, but in general, the flatter the slope of a storm sewer, the greater is the possibility of an error in construction.

The capacity of this segment was determined to be 8.0 cfs from Chart 36 in Appendix A-4 and then entered in Column 12 of Figure 5-25. Since this capacity exceeds the design discharge of 2.6 cfs, the segment was accepted.

The design velocity was determined from Chart 36 to be 5.8 ft/s and entered in Column 13 of Figure 5-25. Time of flow in Segment 1-2 was then computed using velocity and distance to be 0.1 minute and entered in Column 14 of Figure 5-25.

DRAINAGE CALCULATIONS

25-year storm
n = 0.015

(1) Pipe Segment From	To	(2) A—Increm. Area (acres)	(3) C—Runoff Coefficient	(4) A·c—Increm.	(5) A·c—Cummul.	(6) tc—Time of Conc. (min.)	(7) I—Rainfall Intensity (in/h)	(8) Q_p—Peak Runoff (cfs)	(9) Pipe Length (ft)	(10) Slope (%)	(11) Size (in)	(12) Capacity (full) (cfs)	(13) Velocity (fps) (Design Flow)	(14) Travel Time in Pipe (min.)
Southerly System														
1	2	.98	.43	.42	.42	14.5	6.1	2.6	30	2.0	15	8.0	5.8	0.1
2	4	.83	.47	.39	.81	14.6	6.1	4.9	250	8.8	15	17.0	12.0	0.3
3	4	.74	.53	.39	.39	12.2	6.7	2.6	30	2.0	15	8.0	5.8	0.1
4	5	1.12	.39	.44	1.64	14.9	6.0	9.8	50	1.6	18	11.5	7.3	0.1
5	6	.64	.29	.19	1.83	15.0	6.0	11.0	40	1.6	18	11.5	7.4	—
Road 'B'														
7	8	1.70	.46	.78	.78	16.9	5.8	4.5	200	1.0	15	5.5	5.0	0.3
Northerly System														
9	10	1.08	.45	.49	.49	15.4	6.0	2.9	30	2.0	15	8.0	6.0	0.1
10	12	.46	.48	.22	.71	15.5	6.0	4.3	260	3.8	15	11.0	8.2	0.5
11	12	.12	.65	.08	.08	6.0	8.5	0.7	30	2.0	15	8.0	4.0	—
12	13	.64	.50	.32	1.11	16.0	5.9	6.5	125	3.8	15	11.0	9.1	0.2
13	15	.34	.48	.16	1.27	16.2	5.9	7.5	170	3.8	15	11.0	9.5	0.3
14	15	.31	.63	.20	.20	6.0	8.5	1.7	80	2.0	15	8.0	5.0	—
15	16	.52	.51	.27	1.74	16.5	5.8	10.1	40	3.8	15	11.0	10.0	0.1
16	17	—	—	—	1.74	16.6	5.8	10.1	50	1.0	21	14.0	6.3	—

Figure 5-25 Drainage calculation chart for Tall Pines subdivision.

Pipe Segment 2-4

Column 5 of Figure 5-25 shows cumulative Ac to be 0.81, which is the sum of the first two incremental Ac values. Time of concentration is 14.6 minutes based upon the hydraulic path starting in Incremental Area 1 and continuing through Pipe Segment 1-2. Thus $t_c = 14.5 + 0.1 = 14.6$ minutes. This value was used since it is greater than the time of concentration for Incremental Area 2 previously computed to be 13.0 min. Rainfall intensity, based upon a duration of 14.6 minutes, was found in Appendix C-3 to be 6.1 in/h. Therefore, Q_p was computed to be 4.9 cfs and recorded in column 8.

Pipe Segment 2-4 is shown on the profile in Figure 5-21 to be a 15-inch pipe with a length of 250 feet and a slope of 8.8%. Inlet 4 is located at station 0 + 75, which is a shift from the first trial location at station 0 + 50 in order to fit the pipe under the vertical curve and maintain sufficient cover. The capacity was determined to be 17.0 cfs using Chart 36 in Appendix A-4. Since the capacity is greater than Q_p, the pipe segment was accepted.

The design velocity was determined from Chart 36 to be 12 fps. Although this velocity is near the high end of acceptable values, it was not considered problematic since the horizontal deflection between Segment 2-4 and Segment 4-5 is relatively small, thus presenting little probability of surcharging Inlet 4.

Pipe Segment 3-4

This segment is a branch of the system and a cross drain. Peak discharge is computed in columns 2 through 8 to be 2.6 cfs. The size and slope were chosen, as with Segment 1-2, to be 15-inch and 2.0%, respectively. The resulting capacity of 8.0 cfs is greater than Q_p and thus the segment was accepted.

Pipe Segment 4-5

Discharge in this segment consists of all upstream pipes plus runoff directly entering Inlet 4. Therefore, cumulative Ac shown in column 5 of Figure 5-25 is 1.64, which represents the sum of the first four incremental Ac values. Time of concentration is 14.9 minutes based upon the hydraulic path starting in Incremental Area 1 and continuing through Pipe Segments 1-2 and 2-4. All other hydraulic paths yield smaller times of concentration. In Figure 5-25, the value 14.9 minutes is obtained by adding the travel time of 0.3 minutes in column 14 to the time of concentration of 14.6 minutes in column 6 for Segment 2-4. Rainfall intensity of 6.0 in/h was found in Appendix C-3 for a duration of 14.9 minutes. Therefore, Q_p was computed to be 9.8 cfs and entered in column 8.

Pipe Segment 4-5 is shown on the profile to have a length of 50 feet and a slope of 1.6%. The capacity first was determined for a 15-inch pipe to be 7.1 cfs using Chart 36 in Appendix A-4. Since this capacity is less than Q_p, the next higher pipe size of 18 inches was considered. The capacity of an 18-inch pipe was determined to be 11.5 cfs using Chart 37 in Appendix A-4. Since this capacity is greater than Q_p, the segment was accepted as an 18-inch pipe.

Pipe Segment 5-6

Like Segment 4-5 before it, this segment carries discharge from all upstream incremental areas plus runoff entering Inlet 5. Therefore, the cumulative Ac of 1.83 represents the sum of the five incremental Ac values in the southerly system. Time of concentration is 15.0 minutes, which was computed by adding the 0.1 minute travel time in Segment 4-5 to the previous t_c of 14.9 minutes. All other hydraulic paths yield smaller times of concentration. Rainfall intensity of 6.0 in/h was found in Appendix C-3 for a duration of 15.0 minutes. Therefore, Q_p was computed to be 11.0 cfs and entered in column 8.

Pipe Segment 5-6 is shown on the profile to be an 18-inch pipe with a length of 40 feet and a slope of 1.6%. The capacity was determined to be 11.5 cfs using Chart 37 in Appendix A-4. Since this capacity is greater than Q_p, the segment was accepted.

At Manhole 6, the southerly system connects to the existing 24-inch storm sewer in Tall Pines Road. The proposed 18-inch pipe should connect at an elevation in which the crowns of the two pipes match. This then became a vertical constraint on the southerly system and helped determine the 1.6% grade of the last two segments.

Road B

Pipe Segment 7-8

Peak runoff tributary to this segment was computed to be 4.5 cfs as shown in columns 2 through 8 in Figure 5-25. The inlet grate was considered adequate to pass 4.5 cfs discharge because it is located at a low point where a few inches of head can build up above the grate, thus increasing its capacity.

The profile in Figure 5-23 shows a 15-inch pipe with a length of 100 feet and a slope of 1.0%. The 100-foot length was needed to get past the proposed house before beginning the swale, and the slope of 1.0% was dictated by the existing grade as shown on the profile. The capacity was found to be 5.5 cfs using Chart 36 in Appendix A-4. Since the capacity is greater than design discharge, the pipe segment was accepted.

Drainage Swale

Pipe Segment 7-8 discharges into a swale which runs to the detention basin. As shown on the profile in Figure 5-23, the swale starts with a slope of 1.0% which then increases to 2.1% before reaching the detention basin.

To design the swale, first the peak discharge, Q_p, was computed. To compute Q_p, the tributary drainage area was delineated as shown in Figure 5-24. Normally, in designing a drainage swale, the point of analysis is taken as the downstream end of the swale because this gives the most conservative result. However, in this case, as shown in Figure 5-24, due to the peculiar slope of the land the tributary drainage area extends only about 150 feet downstream of the pipe outfall. So this was taken as the point of analysis for computation of Q_p.

Tributary drainage area, A, was measured to be 2.36 acres, which includes the 1.70 acres tributary to Pipe Segment 7-8. Composite runoff coefficient was computed to be 0.40 based upon the following quantities:

Impervious area   = 0.51 acres
Lawn area   = 1.17 acres
Wooded area   = 0.68 acres

Time of concentration was computed based upon the hydraulic path to inlet 7 plus Pipe Segment 7-8 plus 150 feet of the swale:

t_c to inlet 7 = 16.9 min
travel time in segment 7-8 = 0.3 min
travel time in first 150 ft of swale = 1.6 min
   (based upon Figure 4-10, unpaved)

Total t_c = 16.9 + 0.3 + 1.6 = 18.8 min.

Rainfall intensity was taken from Appendix C-3 for a duration 18.8 minutes:

i = 5.4 in/h.

Therefore, Q_p was computed using Equation 4-2:

Q_p = Aci
Q_p = (2.36) (.40) (5.4)
Q_p = 5.1 cfs.

When Q_p was known, the next step was to select a trial cross section and test it for depth and velocity. For trial 1, a trapezoidal channel with a 4-foot bottom width was chosen but was found to be larger than necessary. For trial 2, a triangular channel with side slopes of 2.5 horizontal to 1 vertical was selected.

To determine normal depth and velocity, the trial and error procedure outlined in Section 3.3 was used. An n-value of 0.10 was taken from Appendix A-1 based upon Highway channels and swales with maintained vegetation, fair stand any grass, length 12 inches, depth 0.7 feet—1.5 feet, 2 fps.

This n-value was determined only after a trial and error process to determine approximate depth of flow and velocity.

The resulting depth of flow was 1.4 feet and velocity was 1.1 fps as shown in Table 5-4.

Table 5-4
Computation of Normal Depth in First Section of
Drainage Swale in Tall Pines Subdivision

$$\frac{1.49}{n} s^{1/2} = 1.49$$

D (ft)	a (ft²)	p (ft)	R (ft)	R²ᐟ³ —	v (fps)	Q (cfs)
1.0	2.5	5.4	.463	.598	.64	1.6
2.0	10.0	10.8	.93	.95	1.0	10.0
1.5	5.63	8.1	.69	.78	1.2	6.6
1.4	4.9	7.54	.65	.75	1.1	5.5

Normal depth and velocity were also checked for the steeper portion of the swale where the slope is 2.1%. The resulting depth of flow was 1.2 feet and velocity was 1.5 fps as shown in Table 5-5.

Table 5-5
Computation of Normal Depth in Steeper Section of
Drainage Swale in Tall Pines Subdivision

$$\frac{1.49}{n} s^{1/2} = 2.16$$

D (ft)	a (ft²)	p (ft)	R (ft)	R²/³ —	v (fps)	Q (cfs)
1.4	4.9	7.54	.65	.75	1.6	7.9
1.0	2.5	5.4	.463	.598	1.3	3.2
1.2	3.6	6.5	.557	.677	1.5	5.3

Because the maximum design depth of 1.4 feet remains within the swale with adequate freeboard, and the maximum design velocity of 1.5 fps is non-eroding, the proposed triangular swale was accepted.

Outfall Erosion Protection
To determine whether a riprap outfall apron is needed the design velocity in Pipe Segment 7-8 was compared to the values in Appendix A-2. As shown in column 13 in Figure 5-25, the velocity is 5.0 fps.

In Appendix A-2, the allowable velocity for *silt loam* is 3.0 fps. The choice of silt loam is based upon observations made during site reconnaissance.

Therefore, since the pipe velocity exceeds the allowable velocity, a protective lining is needed and riprap was chosen as the lining material.

First, stone size was determined using Equation 5-1. Tailwater depth was found in Chart 36, Appendix A-4 for a 15-inch pipe with slope 1.0%, n-value 0.015 and discharge 4.5 cfs to be 0.85 feet:

$$d_{50} = \frac{.02}{TW} \left(\frac{Q}{D_0} \right)^{4/3}$$

$$d_{50} = \frac{.02}{.85} \left(\frac{4.5}{1.25} \right)^{4/3}$$

$$d_{50} = 0.13 \text{ ft} = 1.6 \text{ inches.}$$

Although the computed size was 1.6 inches, 4-inch stone was chosen because it is more stable and has little cost difference.

Next, the apron length was computed using Equation 5-2 since TW is greater than $\frac{1}{2}D_0$:

$$L_a = \frac{3Q}{D_0^{3/2}}$$

$$L_a = \frac{(3)(4.5)}{1.25^{3/2}}$$

$$L_a = 9.7 \text{ ft.}$$

Therefore, an apron length of 10.0 feet was chosen as a rounding up of the computed length. The height of the riprap was set 1.0 foot above the tailwater depth, or 0.85 + 1.0 = 1.85 feet. Figure 5-26 shows the riprap design.

Figure 5-26   Riprap outfall protection at headwall 8.

Northerly System

Pipe Segment 9-10
This segment is a cross drain and is designed much like Segment 1-2. Peak discharge was computed to be 2.9 cfs as shown in columns 2 through 8 of Figure 5-25. The size and slope were chosen to be 15-inch and 2.0%, respectively, which results in a capacity of 8.0 cfs.

Pipe Segment 10-12
Peak discharge was computed in a manner similar to segment 2-4. Cumulative Ac is the sum of the first two incremental Ac values, and t_c is the sum of t_c for incremental area 9 plus travel time in Segment 9-10. Therefore, Q_p was computed to be 4.3 cfs.

Pipe Segment 10-12 is shown on the profile in Figure 5-21 to be a 15-inch pipe with a slope of 3.8%. Using Chart 36 in Appendix A-4, the capacity was found to be 11.0 cfs, which is greater than the computed design discharge.

Pipe Segment 11-12
Like Segment 3-4, this is a branch of the system and a cross drain and therefore determined to be a 15-inch pipe with a slope of 2.0%.

Pipe Segment 12-13
Discharge in this segment consists of flow from all upstream pipes plus runoff directly entering Inlet 12. Therefore, cumulative Ac shown in column 5 of Figure 5-25 is 1.11, which represents the sum of the first four incremental Ac values. Time of concentration is 16.0 minutes which was computed by adding the travel time (column 14) in Segment 10-12 to the time of concentration to Inlet 10. (15.5 min + 0.5 min = 16.0 min.) All other hydraulic paths yield smaller times of concentration. Rainfall intensity of 5.9 in/h was found in Appendix C-3 for a duration of 16.0 minutes. Therefore, Q_p was computed to be 6.5 cfs.

Pipe Segment 12-13 is shown on the profile to have a length of 125 feet and a slope of 3.8%. Capacity for a 15-inch pipe was determined to be 11.0 cfs using Chart 36. Since this capacity is greater than Q_p, the segment was accepted.

Pipe Segment 13-15
This segment continues the discharge from the previous segment with the addition of runoff directly entering Inlet 13. Therefore, cumulative Ac shown in column 5 to be 1.27 is the sum of the previous value of 1.11 and the incremental Ac value of 0.16. Similarly, time of concentration shown in column 6 to be 16.2 minutes is the sum of the previous value of 16.0 minutes and the travel time of 0.2 minutes shown in column 14. Rainfall intensity of 5.9 in/h was found in Appendix C-3 just as in the previous segment. Therefore, Q_p was computed to be 7.5 cfs.

The profile shows Segment 13-15 to have a slope of 3.8%, thus giving it the same capacity as the previous segment, namely 11.0 cfs. Since this capacity is greater than 7.5 cfs, the segment was accepted.

Pipe Segment 14-15
Inlet 14 was placed at the low point of the cul-de-sac to prevent puddling. The resulting Pipe Segment 14-15 is a branch of the system and is designed much like Segment 11-12.

Pipe Segment 15-16
This segment conveys discharge from all upstream segments plus runoff directly entering Inlet 15. Cumulative Ac, shown in column 5 to be 1.74, is the sum of cumulative Ac values for Segments 13-15 and 14-15 and incremental Ac for inlet 15. (1.27 + 0.20 + 0.27 = 1.74.) This value also represents the sum of all incremental values shown in column

4. Time of concentration, shown as 16.5 minutes, is computed by adding the travel time in Segment 13-15 of 0.3 minute to the time of concentration to Inlet 13 of 16.2 minutes. (16.2 min + 0.3 min = 16.5 min.) Rainfall intensity of 5.8 in/h was found in Appendix C-3 using a duration of 16.5 minutes. Therefore, Q_p was computed to be 10.1 cfs.

The slope of Segment 15-16 is shown on the profile to be 3.8% thus giving it a capacity of 11.0 cfs; the segment was accepted.

Manhole 16 was included in the layout in order to reduce the slope of the last pipe segment so that the velocity at the outfall to the detention basin could be kept under control. The profile in Figure 5-25 shows that a drop occurs at Manhole 16 allowing Pipe Segment 16-17 to have a 1.0% slope. If the manhole was not included, the pipe would have a slope of approximately 6%, resulting in a velocity of about 13 fps which would be excessive at the outfall.

Pipe Segment 16-17

Since no runoff can enter through the top of the manhole, it has no tributary drainage area, and consequently, columns 2, 3 and 4 of Figure 5-25 are blank. However, the value 1.74 was entered in column 5 because all upstream incremental areas are tributary to Manhole 16 just as they are to Inlet 15. Time of concentration is shown in column 6 to be 16.6 minutes, which is the sum of the time of concentration to Inlet 15 and the travel time in Pipe Segment 15-16. Rainfall intensity of 5.8 in/h was found in Appendix C-3 using a duration of 16.6 minutes. Therefore, Q_p was computed to be 10.1 cfs.

As stated above, the slope of Segment 16-17 was set at 1.0%. At this slope, a 15-inch pipe has a capacity of 5.5 cfs, which is less than the design discharge. Therefore capacity was checked for an 18-inch pipe using Chart 37 in Appendix A-4 and found to be 9.2 cfs, which is still less than the design discharge. Next, a 21-inch pipe was considered and capacity found in Chart 38 to be 14.0 cfs. Also using Chart 38, design velocity for 10.1 cfs was found to be 6.3 fps. Therefore, since its capacity is greater than design discharge and its velocity is reasonably low, the segment was accepted.

It should be noted that a 21-inch RCP is an uncommon size and may not be readily available. Therefore, a 24-inch RCP, which is a more common size, could have been specified in this case since it would perform as well and vertical clearance is not a factor.

The structure chosen for Outfall 17 is a flared end section because the gentle slope of the detention basin fits well against the slope of the flared end section and therefore no wingwalls are needed. Figure 5-14 shows a detail of a flared end section. The apron lining for Outfall 17 was designed as part of the detention basin.

PROBLEMS

1. A 24-inch RCP outfall discharges directly onto the ground. The pipe has a slope of 1.75% and n-value of 0.015 and carries a design discharge of 20. cfs. Design a riprap apron for the outfall.

2. A 60-inch RCP outfall discharges into an earth channel with a bottom width of 5.0 feet and side slopes of 2 horizontal to 1 vertical. The pipe has a slope of 0.8% and n-value of 0.012 and carries a design discharge of 85 cfs. Design a riprap apron for the outfall.

3. A concrete box storm sewer 4 feet high by 8 feet wide discharges into an earth channel with a bottom width of 10 feet and side slopes of 3 horizontal to 1 vertical. The storm sewer has a slope of 0.5% and n-value of 0.015 and carries a design discharge of 215 cfs. Design a riprap apron for the outfall.

4. A 48-inch RCP outfall discharges directly onto the ground. The pipe has a slope of 2.0% and n-value of 0.012 and carries a design discharge of 75 cfs. Design a riprap apron for the outfall.

5. A plan view of a small storm sewer system is shown in Figure 5-27. The site is located in Atlanta, Georgia. Using a design chart like that in Figure 5-25, complete the hydraulic design. Key parameters are as follows:

Design storm: 25-year
Pipe material: reinforced concrete
Manning's n: 0.012
Min. pipe size: 12 inches

Segment	Incremental Area (acres)	c	t_c (min)	Pipe Slope (%)
1-2	1.24	.46	18	1.8
2-3	0.86	.62	16	0.9
3-4	0.52	.80	6	0.5

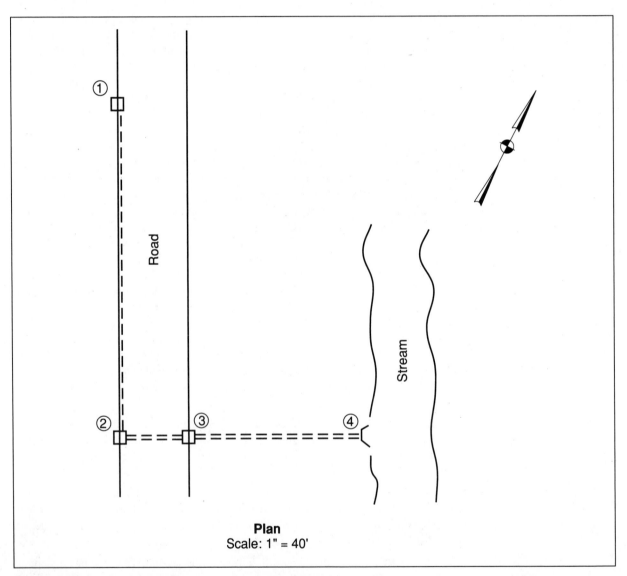

Figure 5-27 Plan of storm sewer system for Problem 5.

6. A site plan for a proposed warehouse located near Pittsburgh, PA (Region 1) is shown in Figure 5-28. Inlet locations and incremental drainage areas have already been determined. The on-site storm sewer system connects through an easement to the existing system in Brickel Avenue. Using a design chart like that in Figure 5-25, complete the hydraulic design for this storm sewer system. Key parameters are as follows:

Design storm:	15-year
Pipe material:	reinforced concrete
Manning's n:	0.012
Min. pipe size:	12 inches

Segment	Incremental Area	c	t_c	Pipe Slope
1-2	0.65	.62	16	2.5
2-5	0.34	.34	15	2.1
3-4	1.16	.28	18.5	1.1
4-5	1.79	.26	19.0	2.4
5-10	—	—	—	1.0
6-7	0.31	.89	6	2.8
7-10	0.15	.65	12	3.5
8-9	0.29	.90	6	1.0
9-10	1.10	.66	10	2.2
10-11	—	—	—	3.0
11-12	—	—	—	1.0

Note: All paved areas are curbed. Roof drains connect to inlets 1, 3 and 4.

7. Design a storm sewer system for Whitebirch Court and Warren Place located in northern New Jersey, which are shown in plan in Figure 5-29. The storm sewer connects to the existing sewer in McDonald Drive at the manhole shown at the intersection. Your design should include drainage area delineation and hydraulic design using a chart like the one in Figure 5-25. System layout is shown in Figure 5-29. Assume all roads are curbed and crowned. Also, assume no runoff enters Warren Place from the road to its south. Wooded areas are delineated with a symbol for wood lines. Suggested design parameters and pipe slopes are shown below.

Design storm:	25-year	
Pipe material:	reinforced concrete	
Manning's n:	0.012	
Min. pipe size:	12 inches	
Runoff coefficients:	paved	0.90
	grass	0.35
	wooded	0.25

Pipe Segment	Slope (%)
1-3	1.5
2-3	2.5
3-5	1.0
4-5	1.0
5-6	6.0
6-7	1.2

Figure 5-28 Site plan showing drainage areas tributary to the inlets.

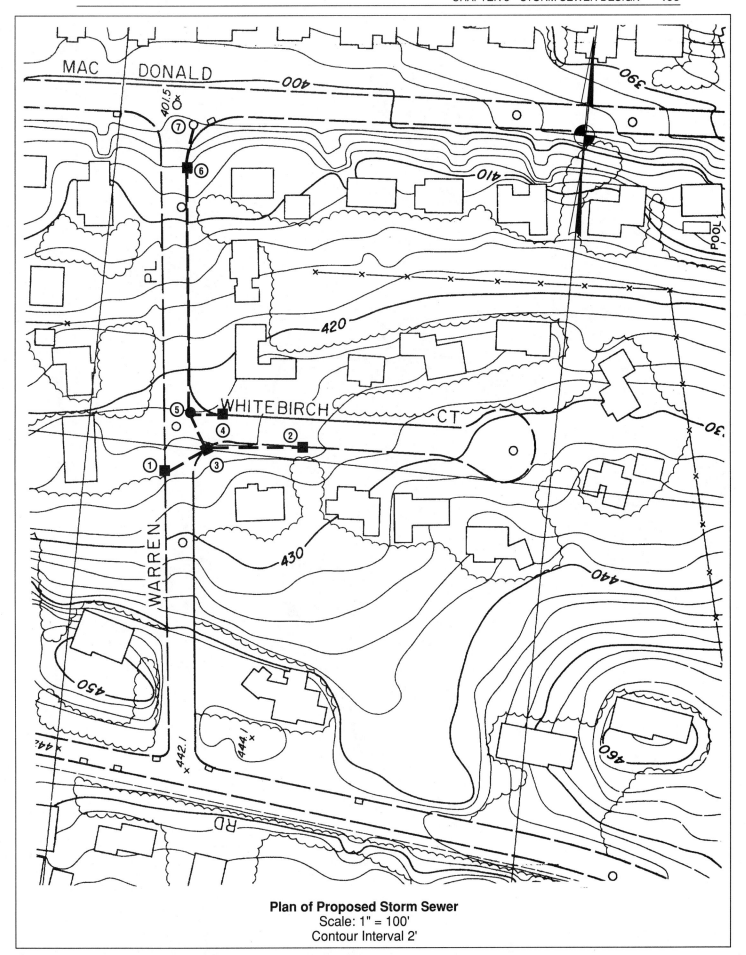

Plan of Proposed Storm Sewer
Scale: 1" = 100'
Contour Interval 2'

Figure 5-29   Plan of proposed storm sewer for Problem 7.

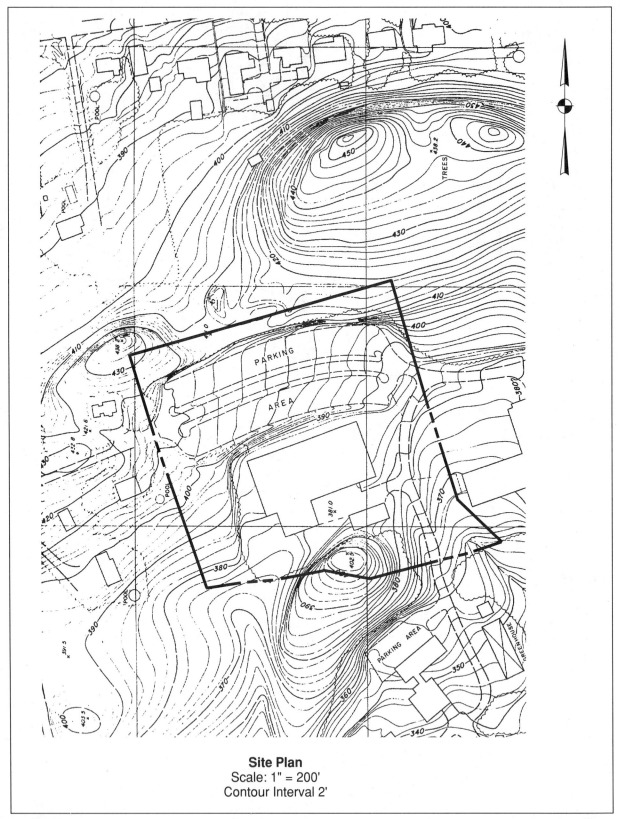

Site Plan
Scale: 1" = 200'
Contour Interval 2'

Figure 5-30 Topgraphic map of developed site located near Atlanta, Georgia. *(Map adapted from Aero Science.)*

8. Design a storm sewer system for the site development located near Atlanta, Georgia and shown in Figure 5-30. The system should outfall at the headwall shown in the southeast corner of the tract. Your design should include system layout, drainage area delineation, hydraulic design and outfall apron design. Assume all paved areas are curbed. Wooded areas are delineated with wood lines. Suggested design parameters are listed below.

Design storm:	25-year		
Pipe material:	reinforced concrete		
Manning's n:	0.012		
Min. pipe size:	12 inches		
Runoff coefficients:	impervious	0.90	
	grass	0.32	
	wooded	0.25	

FURTHER READING

Debo, T., Reese, A. (1995). *Municipal Storm Water Management*. Boca Raton: Lewis Publishers.

Joint Committee of ASCE and Water Pollution Control Federation. (1969). *Design and Construction of Sanitary and Storm Sewers*. New York, NY: ASCE & WPCF.

Urban Water Resources Research Council of ASCE and the Water Environment Federation, (1992). *Design and Construction of Urban Stormwater Management Systems*. New York, NY and Alexandria, VA: ASCE.

6

CULVERT DESIGN

Methods for calculating flow through a culvert were presented in Chapter 3. In this chapter, we will see how to apply those calculations to the design of actual culverts in typical field applications.

As discussed in Chapter 1, many factors must be considered in performing any engineering design and the design principles outlined here for culverts illustrates that assertion. However, it should be remembered that the principle emphasis is on hydraulic design. There are other factors such as structural stability, cost, and traffic control, which will not be treated with as much detail.

OBJECTIVES

After completing this chapter, the reader should be able to:
- Assess an existing culvert for hydraulic adequacy
- Interpret plans showing a culvert profile
- Choose an appropriate layout for a new culvert
- Choose an appropriate layout for a culvert replacement
- Design a culvert size for a new embankment or a culvert replacement
- Design riprap protection for a culvert inlet and outlet
- Relate culvert design to an actual case study

6.1 FUNDAMENTAL CONCEPTS

Design considerations focus on two general concepts: *outcome* and *construction*. Outcome includes all factors determining the finished product while construction involves factors controlling the actual building process.

The principle goal of outcome design for a culvert is determination of the size and alignment of the culvert and provisions for erosion control. The term *size* refers to the dimensions of the barrel, also called the *opening* of the culvert while alignment is concerned with orientation, primarily horizontal.

Additional goals of outcome design include structural stability, durability, cost, ease of maintenance and safety. Thus, the designer visualizes the culvert in its completed, functioning state and seeks to insure that the completed structure will have sufficient hydraulic capacity, sufficient strength, ability to be maintained and all the other desired attributes at an affordable cost.

Principle design goals for the construction phase of the project include erosion control of disturbed soil, control of traffic, safety of workers, control of flooding that may occur during construction, and the routing of heavy trucks through residential areas.

Each culvert design project is unique, but two general categories of project can be identified from the range of individual cases:

1. New Embankment—this involves the design of a new culvert to convey a stream through a new embankment which previously did not exist.

2. Culvert Replacement—this involves the design of a culvert to replace an existing culvert (usually undersized) in an existing embankment.

Each of these design categories is described individually later in this chapter.

The general procedure for hydraulic design is similar for both project categories and is shown in flow chart form in Figure 6-1.

First, the design storm and method of computation must be selected. These are usually specified by the governmental agency having jurisdiction over the culvert. Typically, the design storm is a 100-year storm computed by the SCS Method. However, for minor culverts, a lesser frequency storm may be acceptable.

Allowable headwater elevation (A.H.E.) is a key design parameter and constitutes the maximum height of headwater that will be allowed for the design storm. This height or elevation usually is prescribed by agency regulations and may be the upstream crown of the culvert or may be the lowest point of the top of the embankment or may be some other specific height. If no regulation applies, then the designer must determine the maximum height that the headwater will be allowed to rise during the design storm. The higher the allowable headwater elevation, the smaller the culvert and therefore the more economical the design. Consideration must be given to potential damage caused by the upstream pool of water.

The next preliminary design step is the computation of design discharge. To do this, the watershed must be delineated on a suitable topographic map using the culvert location as the point of analysis. In choosing runoff parameters, the designer must choose between existing watershed ground conditions or conditions of future development.

The more conservative design allows for future development and is therefore preferable, provided no other mitigating circumstances are present. Future development usually means more impervious area and shorter time of concentration, and therefore greater peak runoff. However, with the proliferation of detention basins which are designed to reduce peak runoff, future runoff may not be greater than that occurring at present.

Finally, the water level in the existing stream, which will become the tailwater for the culvert, is determined. If no significant obstructions are present downstream and therefore no backwater effect exists at the culvert location, water level is assumed to be normal depth and is computed using Manning's Equation. However, if backwater is suspected, then water level must be computed using a special method such as standard step calculations. Since this method is beyond the scope of this text, cases involving backwater will not be included. In most cases involving backwater computations, appropriate computer software such as HEC-2 is employed. A list of applicable software currently available is included in Appendix E.

After all the preliminary computations and plottings are complete, the essential hydraulic design may begin. The first step is to choose an initial trial size and shape of the culvert. This includes barrel dimensions, type of wingwalls and entrance geometry. Then compute the headwater elevation using the procedures outlined in Section 3.6. If the resulting headwater elevation is greater than the allowable headwater elevation, another trial must be made using a larger cross section. If the size of the culvert becomes too large to fit under the road, multiple barrels may be used. However, the number of barrels should be kept to a minimum. It is desirable that the opening size of each barrel be as large as possible because small culvert openings are more easily clogged by debris. To find the headwater depth for two culvert barrels, for example, divide design discharge by two and proceed as if for one barrel. Figure 6-2 shows a typical multiple-barrel culvert.

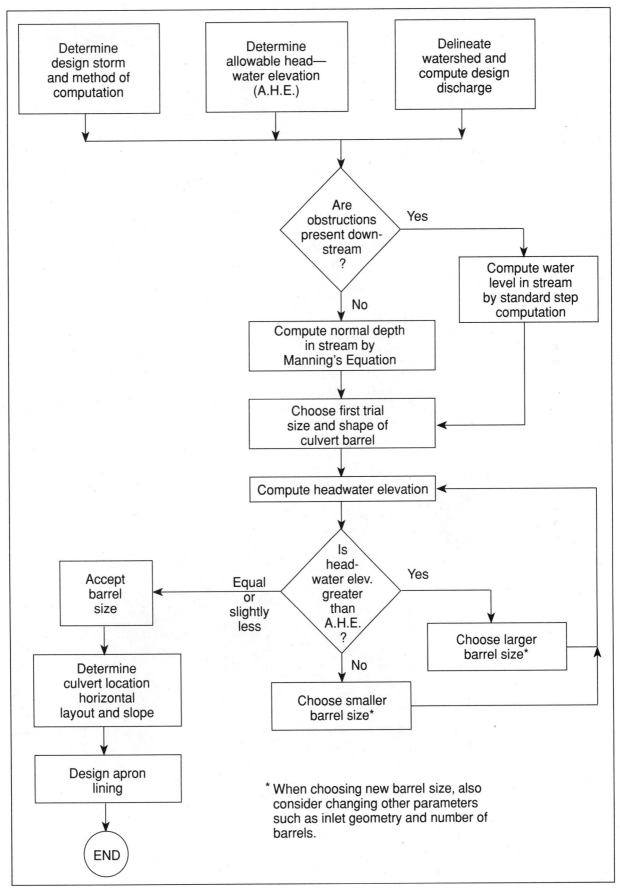

Figure 6-1   Flowchart for culvert hydraulic design procedure.

If the resulting headwater elevation is less than the allowable elevation, the chosen culvert is larger than necessary and consideration should be given to trying a smaller size.

After the barrel opening has been determined, the position or alignment of the culvert can be fixed. In the case of a culvert replacement, the location usually coincides with that of the existing culvert, although there are exceptions. For new culverts, the location is chosen based upon a variety of factors, both horizontal and vertical. Figure 6-3 shows some typical culvert alignments. To assist in vertical alignment, profiles of the stream bed and the top of embankment are plotted.

Generally, the culvert should follow the stream bed in both horizontal and vertical alignment. However, the shortest culvert results from placement perpendicular to the embankment. This may require relocation of a portion of the existing stream as shown in Figure 6-3(a).

Once the culvert size has been established, an apron lining both upstream and downstream of the culvert should be considered. To design an outlet apron lining, compute velocity at the outlet of the culvert by using the continuity equation, Q = va. Cross sectional area is computed by one of three methods as described below.

1. If tailwater depth, TW, is less than critical depth, use critical depth together with barrel dimensions to compute cross-sectional area.
2. If tailwater depth is greater than critical depth, use tailwater depth together with barrel dimensions to compute cross-sectional area.
3. If tailwater level is above the crown of the outlet, use the cross-sectional area of the entire barrel opening.

Using this velocity, design an apron lining in accordance with the procedure outlined in Section 5.5 or any other accepted procedure. Figure 6-4 shows typical riprap lining on the downstream side of a new culvert.

Figure 6-2 Typical three-barrel pipe culvert.

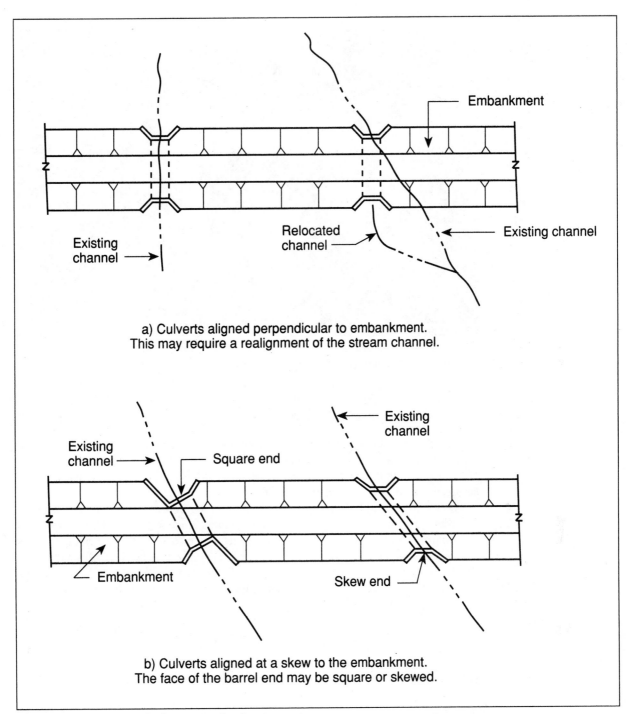

a) Culverts aligned perpendicular to embankment.
This may require a realignment of the stream channel.

b) Culverts aligned at a skew to the embankment.
The face of the barrel end may be square or skewed.

Figure 6-3 Horizontal alignment of culverts.

Figure 6-4 Extensive riprap protects the trapezoidal channel on the down-stream side of this new concrete box culvert.

To design an inlet apron lining, compute velocity at the inlet of the culvert by using the continuity equation. Cross sectional area is computed by using the headwater depth, HW, together with barrel dimensions. If the headwater is above the crown, use the cross-sectional area of the entire barrel opening.

After computing velocity, design the apron lining in accordance with any accepted procedure. For instance, if riprap is to be the lining, stone size may be computed using Equation 5-1 or 5-1a and apron length may be determined at the discretion of the designer.

When computing the cross sectional area of a culvert, if the face of the culvert is skewed, as shown in the right-hand sketch in Figure 6-3(b), do not use the actual skew dimensions. Use the dimensions perpendicular to the long axis of the barrel.

Example 6-1

Problem
Find the outlet velocity for a 4 foot x 8 foot box culvert with a discharge of 200 cfs and tailwater depth of 3.2 feet.

Solution
First, find critical depth by using Chart 7 in Appendix A-3. Critical depth is 2.7 feet.

Since tailwater depth is greater than critical depth, compute cross sectional area using tailwater depth:

$$a = (8)\ (3.2) = 25.6\ \text{ft}^2.$$

Finally, compute velocity using the continuity equation:

$$v = \frac{Q}{a} = \frac{200}{25.6} = 7.8\ \text{ft}/\text{s}.\quad \text{(Answer)}$$

Other environmental considerations are important in culvert design and should be discussed at preliminary meetings with the reviewing agencies. One of these is fish habitat. If the stream is important to fish habitat or production, it may be necessary to design the culvert to protect such conditions. Figure 6-5 depicts two measures designed to protect the fish community. In Figure 6-5(a), a precast concrete culvert is placed lower than the stream invert so that the bed will consist of an unbroken natural channel as it goes through the culvert. In Figure 6-5(b), an arch culvert is placed over a stream to leave the bed and overbank completely undisturbed. Figure 6-6 shows an example of an arch culvert constructed in a new embankment for a road leading to a residential development. The culvert was constructed without touching the stream bed and adjacent wetlands.

Figure 6-5   Examples of methods to preserve existing stream conditions in culverts.

Figure 6-6 Photo of arch culvert spanning a sensitive stream and wetland area.

These environmental considerations have virtually no effect on hydraulic calculations. The only possible effect is the change in roughness due to the presence of stream bed instead of a concrete surface. However, this difference is minor and results in very little discernible difference in headwater elevation.

Another environmental consideration is the potential for flooding downstream from the culvert. In cases of culvert replacement, if the existing culvert is severely undersized it may have been causing significant upstream pooling during severe storms thus reducing the downstream flow rate. In effect, the culvert and embankment act in such cases as a detention basin which attenuates or lessens the flow rate downstream. This effect is analyzed in Chapter 7. Therefore, to replace the undersized culvert with an adequately sized culvert would remove the attenuation and increase the flow downstream, possibly creating flooding problems. If this detrimental effect of increasing the size of a culvert is suspected, a routing analysis should be conducted to assess the downstream flooding problem. The resulting decision to increase the culvert size or to maintain the existing size would typically be a public agency policy decision.

6.2 DESIGN INVESTIGATION

Before beginning a culvert design, certain essential data must be acquired, much as in the case of storm sewer design. These are summarized as follows:

1. Project meeting
2. Relevant maps
3. Site reconnaissance
4. Applicable regulations
5. Related engineering designs or reports

Project Meeting

The project meeting which defines the scope of the design is similar to that described for storm sewer design. It may also be necessary, however, to hold a meeting at the agency which exercises jurisdiction over the proposed culvert.

Culverts generally are considered major hydraulic structures and are directly or indirectly regulated by counties or states or both. In many cases, a special permit or approval is required before constructing a culvert and the application process can be complex. Therefore, a *pre-application* meeting often is needed to clarify how the regulations apply to the project in question and to coordinate between designer and reviewer what design information will be presented with the application.

Relevant Maps

Relevant maps required for design purposes usually consist of a boundary survey and two topographic maps. The boundary survey is needed to be sure that the culvert will be laid out on the owner's property, which usually consists of a road or railroad right-of-way. Certain details of the culvert design, namely the wingwalls and riprap apron, cannot be determined without the use of topography of the culvert area. Also dependent upon this information are certain key vertical relationships such as distance from stream bed to top of embankment. These applications of the topographical data require a high degree of precision, therefore contour lines of 1-foot (0.3 m) or 2-foot (0.6 m) intervals are preferred. In addition, certain spot elevations may be necessary. All existing utilities must also be accurately depicted. Therefore, supplementary field topography may be required.

Topography is also required to delineate the watershed tributary to the culvert. Since the watershed is often quite large compared to the construction site, it may be impractical to use 2-foot (0.6 m) contours, and a smaller scale, less precise topographic map may suffice for runoff calculations. USGS quadrangle sheets drawn to a scale of 1:24000 typically are used for this purpose.

Site Reconnaissance

Next, a site visit must be conducted to verify the topography and make note of any features not clearly depicted on the maps. Also, the stream and overbank must be observed to determine Manning n-values and whether or not obstructions are present downstream.

Regulations and Design Reports

Finally, any applicable regulations must be obtained and reviewed, and if any reports of previous engineering design relating to the project exist, they should also be reviewed.

After all the engineering data have been obtained, the design process may begin.

6.3   DESIGN OF NEW CULVERT

Consider the case of a new road embankment proposed in an alignment that intersects an existing stream. This type of design has some advantages over replacing an existing culvert because there are usually fewer existing constraints, thus allowing more flexibility to the designer. For instance, since the road profile is not yet established, the height of the proposed culvert could possibly be increased and the proposed road raised to compensate. This is not often an option for an existing road.

Another example of increased flexibility is the lack of existing underground utilities such as water, gas and sanitary sewer mains. Also, the absence of improved building lots adjacent to the stream allows increased latitude in horizontal layout of the proposed culvert.

These factors affecting the design of a new culvert in a new road embankment are illustrated by the following example.

Example 6-2

Problem

A new road is proposed to connect from Shawn Court to Oakwood Avenue as shown in Figure 6-7. Design a culvert to convey the existing stream through the proposed embankment.

Solution

The hydraulic design process follows the outline presented in Figure 6-1. Design storm is determined by local regulation to be the 100-year, 24-hour storm as computed by the SCS Method. This computation is then performed and yields a design discharge of 480 cfs.

Allowable headwater elevation (A.H.E.) is determined by local regulation to be the upstream culvert crown.

From the topographic information obtained for the culvert vicinity, a profile of the proposed road centerline is prepared as depicted in Figure 6-8. (At first, the trial culvert openings and road grades are not included; they will be added later.)

Now we may begin hydraulic calculations. The first step is to determine tailwater depth. Since no obstructions are present downstream, flow depth in the stream is computed using Manning's equation. For a design discharge of 480 cfs, tailwater depth is found to be 3.7 feet.

Next, we choose a culvert size as a first trial. Choose a 4 foot x 8 foot concrete box culvert with 45° wingwalls and square edge entrance. To check headwater depth, first assume inlet control. Using Chart 1 of Appendix B-1, Q/B = 61 cfs/ft, which leads to a value of HW/D in scale (1) of the headwater depth scale of 2.9. Therefore, HW = 11.6 feet. This obviously exceeds the A.H.E., which requires a depth of 4 feet at the culvert crown. Therefore, 4 foot x 8 foot is too small, and a larger opening must be tried. Note that it is not necessary to compute HW for outlet control. Why?

For Trial 2, choose a 4 foot x 16 foot culvert. Using Chart 1, HW/D is found to be 1.25, so HW = 5.0 ft, which still exceeds the A.H.E. Therefore, this culvert size is rejected.

For Trial 3, choose a two-barrel culvert with 4 foot x 12 foot openings. To find HW for inlet control, use a design discharge of 240 cfs conveyed by a single 4 foot x 12 foot culvert. Using Chart 1, HW/D is found to be 0.91, which yields a value of HW of 3.64 ft, below the A.H.E. Now, check outlet control.

To find HW assuming outlet control, first find critical depth. From Chart 10 in Appendix A-3, critical depth for 240 cfs is found to be 2.3 feet. Then

$$TW' = \frac{2.3 + 4.0}{2} = 3.15 \text{ ft.}$$

Since TW is greater than TW', use TW for tailwater. Using Chart 8 in Appendix B-2 for k_e = 0.4, and L = 50 ft, H is found to be 0.63 ft. Then,

HW = TW + H − Ls
HW = 3.7 + 0.63 − 0.25
HW = 4.1 ft.

Therefore, the culvert operates under outlet control and the headwater is above the crown. Consequently, this size is rejected.

Figure 6-7   Plan of proposed road location for Example 6-2. *(Map adapted from Aero Service.)*

Figure 6-8   Proposed road profile for Example 6-2.

For Trial 4, choose a two-barrel culvert with 4 foot x 13 foot openings. After computing HW as in the previous trial, the culvert is found to be operating under outlet control with HW = 4.0 feet. Therefore, a twin 4 foot x 13 foot box culvert is hydraulically adequate.

However, other shapes and sizes could be adequate as well provided the resulting head-water is below the crown. For instance, suppose we chose a box culvert with a height of 6 feet. Then for trial 5 you could choose a single 6 foot x 12 foot box culvert. Assuming inlet control, HW is found from Chart 1 to be 5.82 feet. Assuming outlet control, HW is found from Chart 8 to be 4.55 feet. Therefore, this culvert operates under inlet control and HW = 5.82 feet, which is below the A.H.E. So, by the applicable headwater criterium, this culvert also is adequate.

So which culvert is the better choice—twin 4 foot x 13 foot box or single 6 foot x 12 foot box? The latter size would be much less expensive but it would have two disadvantages. First, it would result in a headwater elevation of 227.32, which is 2.12 feet higher than the 100-year flood level occurring without a culvert. For the homes located at the westerly end of Shawn Court, this could be very damaging. Second, the higher culvert would require a higher grade for the proposed road, and therefore much more fill to construct the embankment. Figure 6-8 shows the alternate culvert openings and resulting road profiles.

Therefore, the twin 4 foot x 13 foot box culvert is the better choice because it results in a headwater elevation only 0.3 feet higher than existing and it requires a significantly smaller roadway embankment than would the higher culvert. Figure 6-10 shows a profile of the accepted culvert design.

a) Proposed culvert aligned perpendicular to road.

b) Proposed culvert aligned 45° to the road with square ends.

c) Proposed culvert aligned 25° to the road with skewed ends.

Figure 6-9 Three alternate culvert layouts for Example 6-2. *(Map adapted from Aero Service.)*

If the A.H.E. was determined by allowing no rise in upstream water surface over existing conditions as is required in some locations, the twin 4 foot x 13 foot box culvert would be nearly adequate. In such a case, the smallest adequate culvert would be a twin 4 foot x 14 foot box.

Next, the horizontal layout is determined. Figure 6-9 shows three alternate layouts to fit site conditions. The first alternate, shown in Figure 6-9(a), is the simplest and least expensive. It is perpendicular to the proposed road and has a length of 50 feet. Due to the meandering of the stream, a 160-linear foot channel realignment must be constructed downstream to blend into the existing channel. However, such a modification of the stream may not be allowed by the regulating agency for environmental reasons.

   Note: A linear foot is a unit of horizontal measure which may follow a curved as well as straight line.

The next alternate, shown in Figure 6-9(b), is an attempt to follow the path of the stream to avoid extensive channel realignment. However, even with this layout a small realignment of 60 linear feet is necessary. The amount of skew of the culvert is kept to a minimum because the greater the skew the longer the culvert. (The skew angle is defined as the angle made by the long axis of the culvert and a line perpendicular to the road.) In this case, the culvert is 95 feet long, which is nearly double the length of alternate (a) and increases the cost precipitously. The culvert ends are square because a 45° skew at the end of a culvert is very severe, creating structural problems and adding more cost to the project.

Profile—Proposed Culvert

Figure 6-10 Stream profile for Example 6-2.

The third alternate also attempts to follow the stream path but employs only a 25° skew. In this case, a channel realignment of 50 linear feet is needed at the upstream end to blend with the existing stream. A significant problem with this layout is the orientation of the culvert with the downstream channel. Discharge water is forced to make a sharp left turn as it flows from the culvert into the stream. This could create an erosion problem and also could result in higher tailwater, which could necessitate the increase of the size of the culvert opening.

Another alternative could be to realign the proposed road in a westerly direction to intercept the stream at a straighter section. This would probably require changing the point of intersection on Oakwood Avenue, which may not be possible due to other constraints such as land ownership.

The choice of layout can affect the culvert opening in two ways. First, as mentioned above, different downstream channel conditions lead to different tailwater depths, and second, different barrel lengths can affect headwater depth. Both of these effects apply to culverts operating under outlet control; if the culvert operates under inlet control, layout has little effect on opening size.

For the purpose of this example, we choose layout alternate (a).

Finally, design velocity is computed and the need for any aprons, upstream and downstream, is assessed. Velocity at the outlet is found by using the continuity equation, $Q = va$. Cross-sectional area, a, is the area of a rectangle in which the width is 13.0 feet and the height is 3.7 feet. TW is used for the height instead of D_c because TW is the greater of these values:

$$a = (13.0)\ (3.7) = 48.1 \text{ ft}^2$$

$$v = \frac{Q}{a} = \frac{240}{48.1} = 5.0 \text{ ft}/\text{s}.$$

Referring to Appendix A-2 for earth channels, the permissible velocity for graded loam to gravel is 5.0 ft/s. Since the downstream channel will be constructed for a length of 160 linear feet as part of the project, the soil material can be specified as graded soil, silt to cobbles, and then no additional apron is required.

Velocity at the inlet of the culvert is found in the same manner, except using HW as the height. Thus

$$a = (4.0)\ (13.0) = 52.0 \text{ ft}^2$$

$$v = \frac{Q}{a} = \frac{240}{52} = 4.6 \text{ ft}/\text{s}.$$

Therefore, no apron is required at the upstream end of the culvert as well.

6.4 CULVERT REPLACEMENT

Now, consider the case of an existing road with an existing culvert that is proposed to be replaced. In many culvert replacements, the existing culvert is hydraulically inadequate due to various factors. The culvert could have been constructed at a time when design standards were not as conservative as current standards, development in the watershed could have caused an increase in peak runoff, or re-routing of pipes and channels upstream of the culvert could have increased the tributary drainage area.

The engineer's job in designing a culvert replacement often is made difficult by the physical constraints within which the design must be accomplished. For example, there may be very little vertical clearance from the stream invert to the top of the existing road. This means the culvert cannot be made higher, just wider. In some cases, the road can be raised, but this adds considerable cost to the project.

Another typical constraint is the presence of established improvements and utilities adjacent to the existing culvert and stream. This eliminates the option of changing culvert layout.

In performing design of a culvert replacement, the existing culvert is normally analyzed first for its hydraulic adequacy to determine whether the existing opening must be enlarged. Then determination should be made whether the road is scheduled for future widening so that the new culvert could be designed longer, if necessary.

These design considerations for a culvert replacement project are illustrated in the following example.

Example 6-3

Problem

The existing 36" RCP culvert under Washington Road as depicted in Figure 6-11 is proposed to be replaced. Design a replacement culvert.

Solution

First, the existing culvert is assessed for hydraulic adequacy. If found to be inadequate, a new culvert is designed following the outline presented in Figure 6-1.

Design storm is determined by local regulation to be the 100-year, 24-hour storm as computed by the SCS Method. This computation is then performed and yields a design discharge of 131 cfs.

Allowable headwater elevation (A.H.E.) is determined by local regulation to be the inundation of the road or the elevation at which floodwater will start to run over the road. This elevation is determined from the topography to be 415.0. (Note: even though the low point elevation of the road profile is 411.50, the controlling elevation of 415.0 occurs just west of the culvert on the north side of the road at a low point in the stream bank.

Existing road and culvert profiles are depicted in Figures 6-12(a) and 6-13(a), respectively.

As shown in Figure 6-13(a), the existing culvert consists of a 36" RCP at the upstream end with a 48" CMP finishing at the downstream end. The upstream end has no wingwall while the downstream end is set in an elaborate concrete apron and wingwall structure. Evidence of considerable erosion at the downstream end is manifest by the drop in stream invert at the end of the apron.

Apparently the original 36" RCP culvert was previously extended for road widening by slipping a 48" CMP over the concrete pipe.

To assess the hydraulic adequacy of the existing culvert, first check inlet control. Using Chart 2 in Appendix B-1 for a 36-inch diameter pipe and design discharge of 131 cfs, HW/D is found on scale (3) to be 4.7. Therefore,

$$HW = (4.7)(4.0) = 18.8 \text{ ft}$$

and the headwater elevation is equal to the upstream invert plus HW, or

$$411.7 + 18.8 = 430.5,$$

which clearly exceeds the A.H.E. of 415.0. Therefore, the existing culvert is inadequate.

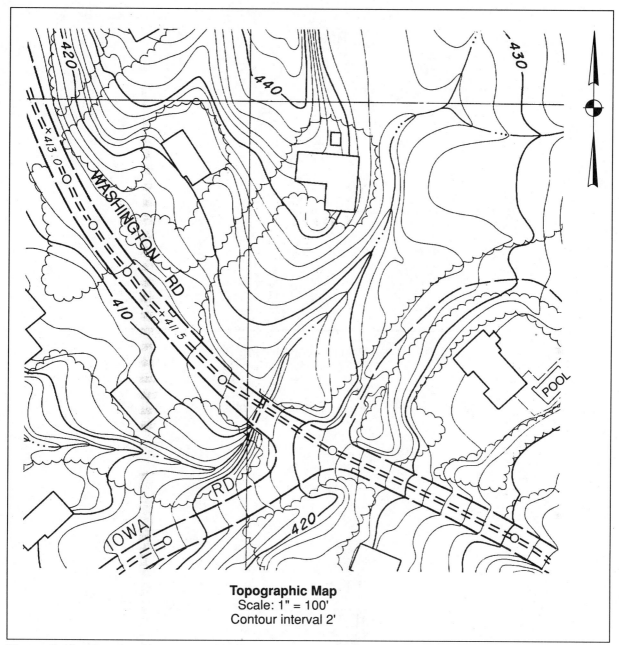

Figure 6-11   Topographic map used in Example 6-3. *(Map adapted from Aero Service.)*

Normally the first step in designing a replacement culvert would be determination of opening size. However, in this case, since vertical alignment can affect capacity, layout will be analyzed first. The horizontal layout will follow the existing culvert due to property constraints. However, as we can see in the profile, Figure 6-13(a), considerable vertical variation is available. Currently, the downstream invert is about 4.5 feet above the stream invert, so the entire culvert could be lowered to match the stream invert. However, this would create a conflict with the existing sanitary sewer main that crosses under the culvert, although the main could be lowered as shown in Figure 6-12(b).

If the new culvert is lowered at the outlet end, it could also be lowered at the inlet end as shown in Figure 6-13(b). This would create the advantage of allowing a headwater pool to develop during the design storm, which would increase the capacity of the

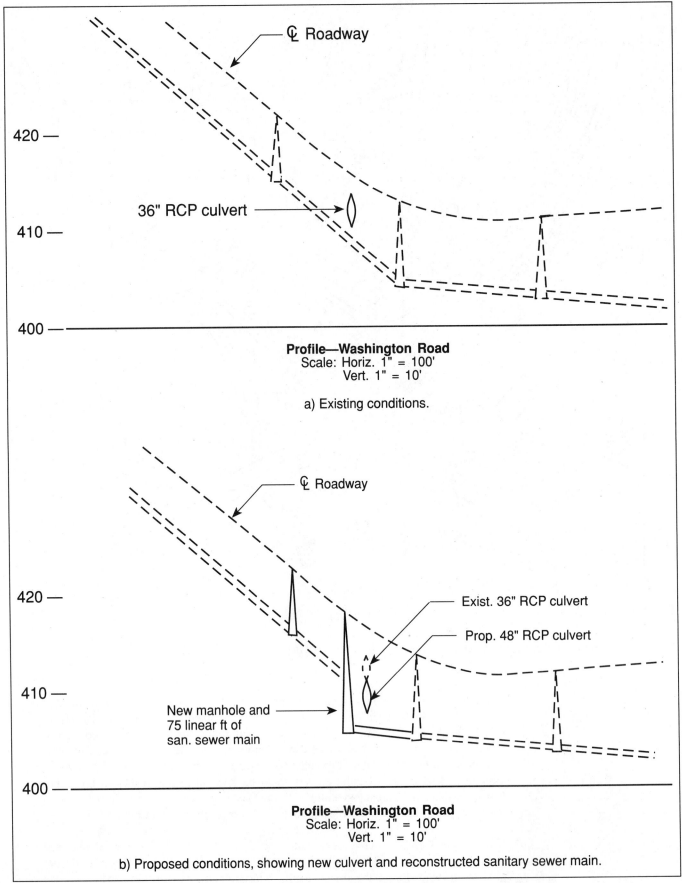

Profile—Washington Road
Scale: Horiz. 1" = 100'
Vert. 1" = 10'

a) Existing conditions.

Profile—Washington Road
Scale: Horiz. 1" = 100'
Vert. 1" = 10'

b) Proposed conditions, showing new culvert and reconstructed sanitary sewer main.

Figure 6-12 Profiles of Washington Road for existing and proposed conditions.

a) Existing conditions.

b) Proposed conditions showing new culvert and lowered stream channel at upstream end.

Figure 6-13   Profiles of existing and proposed culverts for Example 6-3.

culvert. However, this is only true if the culvert operates under inlet control. If outlet control prevails, then lowering the inlet end would have no effect on capacity.

Following this reasoning, choose a 36" RCP at the lower profile shown in Figure 6-13(b) for Trial 1. Let the new culvert have wingwalls and square edge. To check inlet control use Chart 2 in Appendix B-1. For a 36-inch pipe, with entrance type (1),

$$\frac{HW}{D} = 5.6$$

$$HW = (5.6)\,(3.0) = 16.8 \text{ ft}$$

and headwater elevation is

$$407.8 + 16.8 = 424.6.$$

Since this elevation exceeds the A.H.E., Trial 1 is rejected.

For Trial 2, choose a 48" RCP at the lower profile used in Trial 1. To check inlet control, use Chart 2 in Appendix B-1. For a 48-inch pipe with entrance type (1),

$$\frac{HW}{D} = 1.8$$

$$HW = (1.8)\,(4) = 7.2 \text{ ft}$$

and headwater elevation is

$$407.8 + 7.2 = 415.0.$$

Since this elevation equals the AHE, check outlet control.

To check outlet control, first compute tailwater depth, TW. Since the stream has no downstream obstructions, Manning's equation is used. From the topographic map, the gradient of the stream is computed to be 5.0%, and a field observation is used to determine an n-value of 0.04 (mountain stream, no vegetation in channel, steep banks, cobbles on bottom). The channel cross-section (looking downstream) is shown in Figure 6-14. Using Manning's equation, normal depth is found to be 1.25 feet. This depth is taken as TW. Note this very shallow depth is due to the steep slope of the stream.

Next, compute critical depth. Using Chart 45 in Appendix A-4,

$$D_c = 3.4 \text{ ft.}$$

Therefore TW' is computed as

$$TW' = \frac{(D_c + D)}{2}$$

$$TW' = \frac{3.4 + 4}{2} = 3.7 \text{ ft.}$$

420 —

410 —

400

Existing stream cross section
outlet of culvert
(looking downstream)
Scale: 1" = 10'

a) Existing stream cross section.

Riprap, 20" thick
d_{50} = 20"

3'   6'   3'

3' x 3' x 6' gabions
stacked 2 high

2" crushed stone, 6" thick
Filter fabric

Proposed stream cross section
outlet of culvert
(looking downstream)
Scale: 1" = 10'

b) Proposed stream cross section showing gabions and riprap.

Figure 6-14   Cross section of stream at downstream end of culvert for existing and proposed conditions for Example 6-3.

Since TW' > TW, use TW' to compute headwater elevation. Next, determine H using Chart 9 in Appendix B-2. For Q = 131 cfs, k_e = 0.5,

H = 2.9 ft.

Therefore, headwater elevation is

405.4 + 3.7 + 2.9 = 412.0.

Since a higher headwater elevation is computed for inlet control, the culvert operates under inlet control and headwater elevation is 415.0.

Therefore, the culvert defined in Trial 2 is adequate.

As seen in Example 6-2, other culvert arrangements might also be hydraulically adequate. So consider some other trials. The fact that the existing culvert already consists of a 48" CMP at the outlet end suggests the possibility of simply replacing the 36" RCP and extending the 48" CMP up to the inlet end. This alternative would be far less expensive than Trial 2 since the existing wingwalls could be preserved and the sanitary sewer could be left in place. So, Trial 3 is a 48" CMP at the existing elevation. Unfortunately, we can see immediately that if the inlet end is at a higher elevation than Trial 2, the headwater level will exceed the A.H.E. and the trial must be rejected.

Another problem with Trial 3 is that preserving the downstream wingwall would not make allowance for future road widening, which would be accomplished by Trial 2.

For Trial 4, let's consider a 4 foot x 4 foot concrete box culvert at the lower elevation used in Trial 2. To check inlet control, use Chart 1 in Appendix B-1.

$$\frac{Q}{B} = \frac{131}{4} = 32.75 \text{ cfs} / \text{ft.}$$

Thus,

$$\frac{HW}{D} = 1.33$$

$$HW = (1.33)(4.0) = 5.3 \text{ ft.}$$

Therefore, headwater elevation is

$$407.8 + 5.3 = 413.1.$$

To check outlet control, first determine critical depth from Chart 3 in Appendix A-3 or from Equation 3-2:

$$D_c = 3.2 \text{ ft.}$$

Therefore, TW' is computed as

$$TW' = \frac{(D_c + D)}{2}$$

$$TW' = \frac{3.2 + 4}{2} = 3.6 \text{ ft.}$$

Since TW' > TW, use TW' to compute headwater elevation. Next, determine H using Chart 8 in Appendix B-2:

$$H = 1.7 \text{ ft.}$$

Therefore, headwater elevation is

$$405.4 + 3.6 + 1.7 = 410.7.$$

Since a higher headwater elevation is computed for inlet control, the culvert operates under inlet control and headwater elevation is 413.1.

Therefore, the culvert defined in Trial 4 is adequate.

In fact, since the headwater elevation for Trial 4 is 1.9 feet lower than the A.H.E., the inlet end of the culvert could be raised by that amount, thus decreasing the amount of excavation required for installation. However, raising the culvert 1.9 feet to elevation 409.7 would not eliminate the need to reconstruct the sanitary sewer main as in Trial 2.

With all cost factors considered, Trial 4 is significantly more expensive than Trial 2. Therefore, Trial 2 is accepted as the hydraulic design for the culvert replacement project.

Finally, design velocity is computed and the need for any aprons, upstream and downstream, is assessed. Velocity at the outlet is found by using the continuity equation. Cross sectional area, a, is the area of a segment of a circle with diameter 4.0 feet. Depth of flow is critical depth, 3.4 feet, since $D_c > TW$. Using Figure 3-9, 3.4 feet is entered as a percent of the diameter or

$$\frac{3.4}{4.0} = 100\% = 85\%$$

and the corresponding area is read as a percent, or 90% of total area. Therefore,

$$a = (.90)(12.57) = 11.3 \text{ ft}^2,$$

and velocity is

$$v = \frac{Q}{A} = \frac{131}{11.3} = 11.6 \text{ ft/s.}$$

Referring to Appendix A-2, all permissible velocities for earth channels are exceeded. Therefore a riprap lining is designed as an outlet apron. From Equation 5-1

$$d_{50} = \frac{.02}{1.25}\left(\frac{131}{4}\right)^{4/3}$$

$$d_{50} = 1.67 \text{ ft} = 20 \text{ inches.}$$

From Equation 5-2,

$$L_a = \frac{(3)(131)}{4^{3/2}}$$

$$L_a = 49 \text{ ft.} \quad \text{(use 50 ft)}$$

Since a channel exists downstream of the culvert, the riprap will line the channel cross-section to a height of 1.0 foot above the tailwater depth. However, a close look at Figure 6-14(a) reveals very steep side slopes. Riprap should not be placed on a slope steeper than two horizontal to one vertical. Therefore, to accommodate riprap, the downstream channel banks should be regraded with two horizontal : one vertical slopes. Unfortunately, not enough room is available in the channel for such grades without severely cutting into adjacent properties. Therefore, we will propose a combination of riprap on the channel bottom and gabions on the sides running for 50 feet along the channel as shown in Figure 6-14(b).

The riprap thickness is only 20 inches because the relatively large stone size should be stable and a blanket of crushed stone with a filter fabric is provided below the riprap. Next, design velocity is computed at the inlet end of the culvert. Since the culvert operates in a submerged condition, cross sectional area is that of the entire opening, 12.57 ft². Therefore,

$$v = \frac{131}{12.57} = 10.4 \text{ ft/s.}$$

Referring to Appendix A-2, all permissible velocities for earth channels are again exceeded. Therefore, a protective lining is needed for the channel approaching the culvert. Part of the overall culvert design involves a reconstructed stream channel for a distance of 25 linear feet to drop the stream invert to the new lower invert of the culvert. This results in a channel slope over 10 percent which will cause excessive velocities under low flow conditions. That is, the most erosive conditions are not during the 100-year storm but lesser storms which do not cause ponding at the culvert inlet. Riprap design for this condition is beyond the scope of this text.

The best design solution for erosion protection at the upstream side of this culvert is the use of either gabions, grouted riprap or a concrete slab lining.

6.5 ENTRANCE EFFICIENCY

The culvert in Example 6-3 operates under inlet control meaning that during the design storm it is harder for water to get into the culvert than it is to pass through the rest of the barrel and out the end. This means that perhaps a smaller barrel could be used if there was some way to get the water in initially.

Such a problem can be solved by changing the geometry of the culvert entrance. We have already discussed that simply adding a bevel edge to the entrance can somewhat increase entrance capacity. In addition, special inlet designs called *side-tapered* and *slope-tapered* inlets can be employed to significantly increase entrance capacity.

Figure 6-15 shows sketches of the essential features of side tapered and slope tapered inlets. These are concrete structures constructed at the entrance to a culvert and have the effect of reducing entrance head loss. Slope-tapered inlets are the more effective, but they require a vertical distance from the culvert invert to the stream invert which often is not available.

In the case of Example 6-3, however, such vertical space was available making the culvert a good candidate for an improved inlet. But by lowering the stream, a similar effect was realized. If the stream was not lowered, the culvert would have been much larger and therefore more expensive. Nonetheless, the culvert could possibly be reduced to even a smaller size, say 42" RCP, by constructing a combination side tapered and slope tapered inlet.

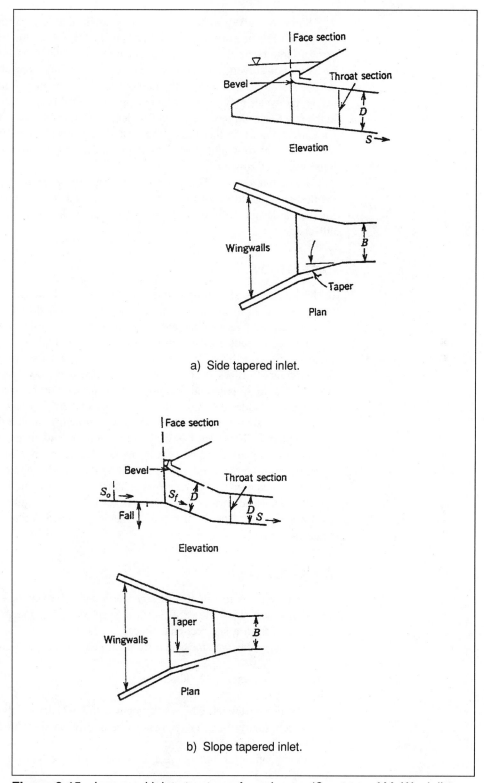

a) Side tapered inlet.

b) Slope tapered inlet.

Figure 6-15 Improved inlet structures for culverts. *(Courtesy of M. Wanielista, Hydrology and Water Quantity Control, John Wiley & Sons, Inc.)*

Since the culvert already is relatively small, the savings realized by going from 48 inches to 42 inches would probably not justify the cost of the improved inlet. But in the case of large box culverts, the savings can be dramatic. Consider, for example, a 6 foot x 16 foot box culvert, 100 feet long operating under inlet control. The size of the opening (6 foot x 16 foot) serves only the purpose of letting enough water into the culvert. Once inside, the water may require a much smaller barrel (say 4 foot x 12 foot) to convey it to the other end. So the entire 100-foot length is sized at 6 foot x 16 foot simply because the entrance must be that large. The cost saving in reducing 100-feet of culvert from 6 foot x 16 foot to 4 foot x 12 foot is probably well worth the cost of constructing an improved inlet structure.

The actual design of improved inlet structures is beyond the scope of this text, but the designer should be aware of this option. Various design manuals include procedures for such a design. Examples of these are included in the Further Reading at the end of this chapter.

6.6 CASE STUDY

In this case, a road is proposed to be reconstructed along a 2,200-linear foot length, including wider pavement, new curbs, storm sewers and two culvert replacements. The road, named Southern Boulevard, is located in eastern Massachusetts.

This case study focuses on one of the culvert replacements called Culvert No. 1, located at Station 182 + 70. The existing culvert is a 4 foot x 4 foot box culvert conveying a tributary to Black Brook as shown in Figure 6-16. Because the road will be widened from 22 feet to 36 feet, and because the existing culvert construction is substandard, the decision was made to completely replace the culvert.

The first step in the design process was to determine the design storm and method of computation. These are based upon the requirements of the regulating and reviewing agency. In this case, the culvert is owned by the county in which it is located and therefore must be reviewed and approved by the county engineering department. County regulations required a 100-year design storm to be computed by the SCS Method.

Allowable headwater elevation was set by county regulations to be no higher than the lowest point of the road profile. That is, the road could not be overtopped by the headwater pool.

Design Discharge

The next step was to calculate design discharge. The watershed tributary to the culvert was delineated using the USGS Quad sheet covering the project area together with information from a site reconnaissance.

Drainage area was measured by planimeter to be 234 acres. Therefore,

$$A_m = 234 \text{ acres} = 0.366 \text{ s.m.}$$

The curve number, CN, was computed to be 71 based upon soils delineated on the county soil survey map, cover conditions shown on the Quad sheet, and observations made during a site visit.

Precipitation for the 100-year, 24-hour storm in eastern Massachusetts was found in Appendix D-3:

$$P = 6.7 \text{ inches.}$$

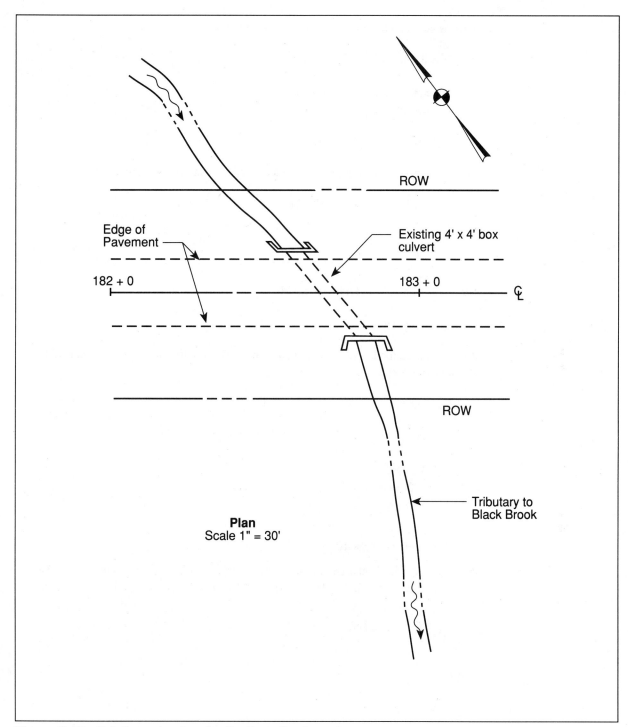

Figure 6-16 Existing culvert under Southern Boulevard.

Next, runoff depth, Q, was determined using Figure 4-20:

Q = 3.5 inches.

Time of concentration was determined to be 1.05 hour.
Rainfall distribution was found in Appendix D-4 to be

Type III.

Next, unit peak discharge, q_u, was determined for a type III rainfall distribution:

I_a = .817   (from Table 4-2)

$$\frac{I_a}{P} = \frac{.817}{6.7} = .12.$$

Using Appendix D-5 (type III rainfall),

q_u = 280 csm/in.

The swamp factor F_p was determined from Table 4-4 for an estimated 2 acres (1%) of swamps and ponds:

F_p = 0.87.

Finally, peak runoff, q_p, was computed using Equation 4-7:

$q_p = q_u A_m Q F_p$
q_p = (280) (.366) (3.5) (.87)
q_p = 312 cfs.

Existing Culvert Adequacy

Now that design discharge was determined, the existing culvert could be checked for hydraulic adequacy. First, inlet control was assumed. Using Chart 1 in Appendix B-1,

$$\frac{Q}{B} = \frac{312}{4} = 78.0 \text{ cfs} / \text{ft.}$$

and

$$\frac{HW}{D} = 4.2$$

$$HW = (4.2) (4.0) = 16.8 \text{ ft.}$$

Therefore, headwater elevation was computed to be equal to the invert elevation at the inlet plus HW, or

244.0 + 16.8 = 260.8

which was far above the low point of the road centerline, which had elevation 249.72 as shown in Figure 6-17. The actual headwater elevation would be much less than the computed 260.8, but greater than 249.72, because as headwater rises above the road, the backwater pool begins to flow over the road as though it is a large weir. Therefore, actual discharge would be a combination of culvert flow and weir flow resulting in a headwater elevation about 8 inches above the road, or approximately 250.4.

This computation established the fact that the existing culvert was hydraulically inadequate since the road would be overtopped. Since the road cannot be overtopped, the allowable headwater elevation A.H.E. became that of the proposed road low point, or 249.42.

Hydraulic Design

With the allowable headwater elevation established, hydraulic design of the proposed culvert could begin. First, tailwater elevation was computed using Manning's equation since the downstream channel contained no significant obstructions creating a backwater condition. A cross section of the stream at the culvert outlet is shown in Figure 6-18. Normal depth was determined to be 4.13 feet giving a tailwater elevation of 247.80, which was above the culvert crown. What does this tell you about inlet control?

For a first trial in selecting culvert size, a box culvert with height 4 feet was assumed since the proposed road elevation would not allow a culvert with greater height. So trial 1 was taken as a 4 foot x 8 foot box culvert.

Figure 6-17   Profile of Southern Boulevard showing proposed culvert.

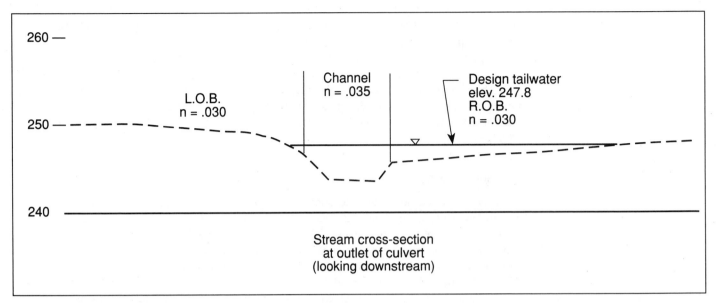

Figure 6-18 Cross section of stream at outlet end of culvert.

Since tailwater was above the crown, outlet control was assumed. Entrance coefficient, k_e, was found in Appendix B-3 to be 0.4 for flared wingwalls and square edge. Then using Chart 8 of Appendix B-2,

H = 2.3 ft.

Therefore, headwater elevation was equal to tailwater elevation plus H, or

247.8 + 2.3 = 250.1.

This elevation exceeds the A.H.E. of 249.42 so Trial 1 was rejected.

For Trial 2, a 4 foot x 10 foot box culvert was assumed. Again using Chart 8 in Appendix B-2,

H = 1.5 ft.

Therefore, headwater elevation was equal to tailwater elevation plus H, or

247.8 + 1.5 = 249.3.

This elevation is less than the A.H.E. so Trial 2 was accepted and the culvert size set at 4 foot x 10 foot.

Layout

The next step was the culvert layout. Because the land outside the road right-of-way was not owned by the county, any redirection of the stream was precluded and the only culvert layout option was a skew similar to the orientation of the existing culvert. Figure 6-19 illustrates the process used to determine the layout.

ROW

182 + 0   183 + 0

ROW

Inset
End section of
culvert with
36°30' skew.

a) First attempt to locate the proposed culvert.

ROW

182 + 0   183 + 0

30" RCP   15" RCP   36" RCP

ROW

Inset
End section of
culvert with
29° skew.

b) Final location of proposed culvert.

Figure 6-19   Location of proposed culvert within road right-of-way.

In Figure 6-19(a), a precast concrete box culvert was laid out with sufficient length to clear the proposed road with extra room for sidewalks on both sides. This resulted in a skew angle of 36°30', which created difficulties in manufacturing the two end sections. A solution was conceived by adding bend sections near the two ends as shown in Figure 6-19(b). Slightly bending the culvert resulted in end sections with 29° skews, which was within acceptable manufacturing limits. The bend sections are illustrated more clearly in Figure 6-20, which shows the final culvert design. The slight bending of the culvert would not significantly alter its hydraulic functioning.

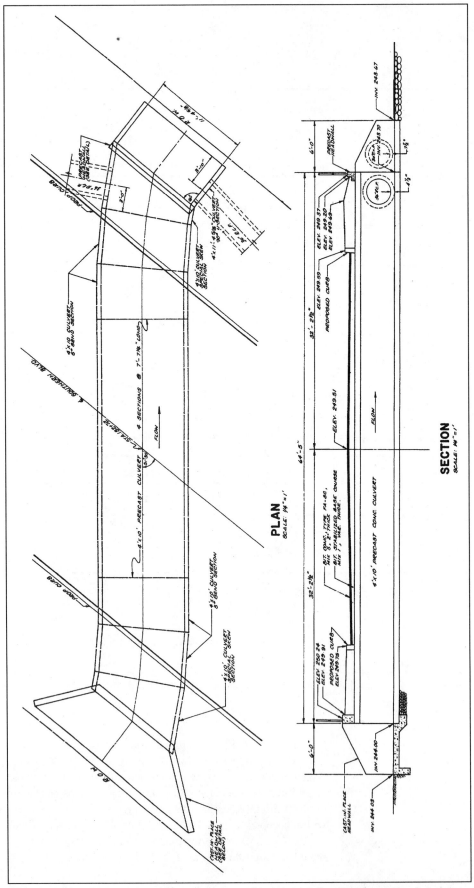

Figure 6-20 Plan and section of proposed culvert under Southern Boulevard. *(Courtesy of Keller and Kirkpatrick.)*

Apron Design

Finally, a riprap lining for the stream was designed. The outlet velocity was computed using the continuity equation, taking the full culvert opening for cross-sectional area because tailwater was above the crown. Therefore,

$$v = \frac{Q}{a}$$

$$v = \frac{312}{40}$$

$$v = 7.8 \text{ ft/s.}$$

This velocity exceeds all permissible velocities included in Appendix A-2, so riprap was chosen as a suitable channel lining. Stone size was computed using Equation 5-1.

$$d_{50} = \frac{.02}{TW}\left(\frac{Q}{D_0}\right)^{4/3}$$

$$d_{50} = \frac{.02}{4.13}\left(\frac{312}{10}\right)^{4/3}$$

$$d_{50} = 0.48 \text{ ft} = 5.7 \text{ inches}$$

use $d_{50} = 6$ inches.

The riprap lining length was computed using Equation 5-2:

$$L_a = \frac{3Q}{D_0^{3/2}}$$

$$L_a = \frac{(3)(315)}{10^{3/2}}$$

$$L_a = 29.9 \text{ ft}$$

use $L_a = 30.0$ ft.

Therefore, riprap was proposed for a length of 30 feet downstream from the culvert using a median stone size of 6 inches and thickness of 12 inches with filter fabric on the bottom. The riprap was proposed to extend to the top of the right bank, which is only 2 feet high, and to the top of the left bank, which is about 3 feet high.

The downstream riprap would be contained within an existing stream easement. On the upstream end of the culvert, no riprap was proposed since no easement existed or could be obtained outside the extent of the road right-of-way. However, as shown in Figure 6-20, an upstream concrete apron was provided.

PROBLEMS

1. A 30-foot wide road is proposed to be constructed over an existing stream as shown in Figure 6-21. Determine the best layout for a proposed 10-foot wide box culvert to fit the following constraints:

   a. The culvert must extend 10 feet beyond each edge of the proposed road.
   b. The cost of the proposed culvert is $2000 per linear foot.
   c. The cost of stream relocation is $250 per linear foot.
   d. The maximum culvert skew is 24°.

   To analyze this problem, place a piece of transparent vellum over Figure 6-21 and trace the plan, then sketch the proposed culvert on the vellum.

2. The opening size of a proposed culvert under a new road is to be determined. The culvert, shown schematically in Figure 6-22, will have a height no greater than 5.1 feet. Design discharge has been computed to be 420 cfs. Find the culvert opening dimensions for the following shapes:

   a. Circular concrete pipe
   b. Concrete box

   Note: Multiple barrels are allowed and each culvert has 45-degree wingwalls and square edge entrance. Which shape would you choose?

3. Figure 6-23 shows an existing road and existing culvert. Evaluate the hydraulic adequacy of the culvert for a design discharge of 860 cfs and A.H.E. equal to the top of road, or elevation 184.5. Computed tailwater depth is 3.9 feet giving a tailwater elevation of 181.3.

   If the culvert is inadequate, design a replacement culvert using flared wingwalls and square edge entrance. Also, design riprap outlet protection if needed. Figure 6-24 shows the existing stream profile.

4. Figure 6-25 shows an existing road and existing culvert consisting of three 30-inch concrete pipes. Evaluate the hydraulic adequacy of the culvert for a design discharge of 152 cfs and A.H.E. equal to the top of road, or elevation 320.9. Computed tailwater depth is 3.25 feet, giving a tailwater elevation of 320.00.

   If the culvert is inadequate, design a replacement culvert using flared wingwalls and square edge entrance. Also design riprap outlet protection if needed. Keep in mind that the culvert crown cannot be made much higher because of the existing road. Figures 6-26 and 6-27 show profiles of Brickel Avenue and the stream respectively.

5. Figure 6-28 shows the location of a proposed road to connect Beth Street to Dwight Street. Design a culvert to convey the existing stream under the proposed road. Design discharge has been computed to be 340 cfs. Figures 6-29 and 6-30 show profiles of the proposed road and existing stream, respectively.

   Notice the water surface profile shown in Figure 6-30. The profile shows a flow depth of 3.25 feet upstream of the proposed culvert and 3.0 feet downstream. The upstream depth is needed to evaluate the A.H.E., which is defined as being no higher than 0.2 foot above existing water level.

   Also, design riprap outlet protection, if needed.

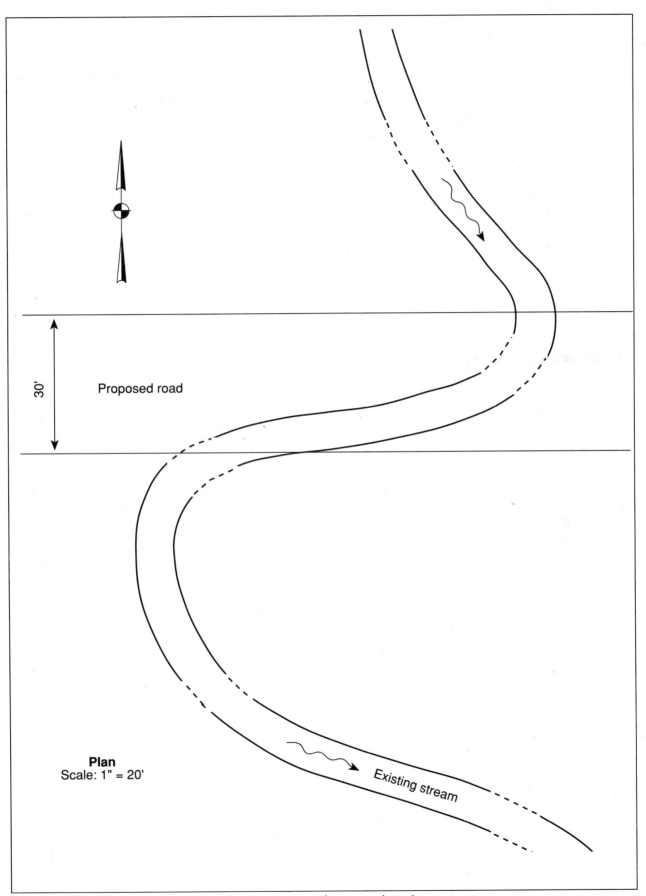

Figure 6-21 Layout plan showing existing stream and proposed road.

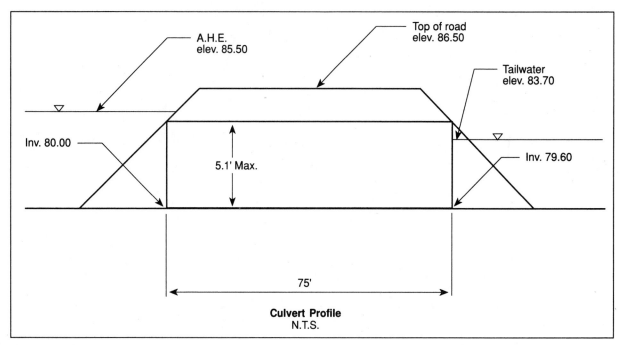

Figure 6-22   Schematic profile of culvert.

Topographic Map
Scale: 1" = 100'
Contour interval 2'

Figure 6-23   Road plan and profile showing existing box culvert. *(Map adapted from Aero Science.)*

Figure 6-24 Stream profile showing existing culvert.

Topographic Map
Scale: 1" = 100'
Contour interval 2'

Figure 6-25   Plan of existing culvert. *(Map adapted from Aero Service.)*

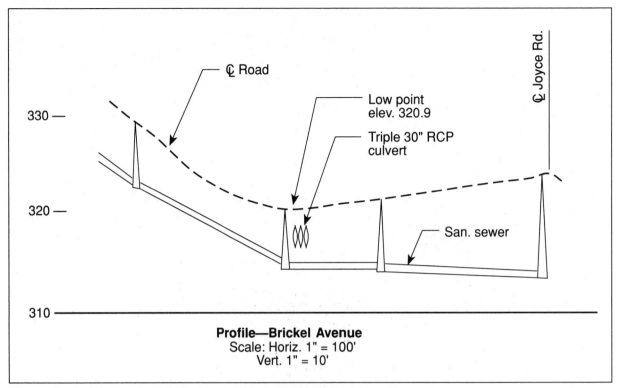

Figure 6-26   Profile of road showing existing culvert for Problem 4.

Figure 6-27   Profile of stream showing existing culvert for Problem 4.

Topographic Map
Scale: 1" = 100'
Contour interval 2'

Figure 6-28 Plan of proposed road location. *(Map adapted from Aero Service.)*

Figure 6-29   Profile of proposed road for Problem 5.

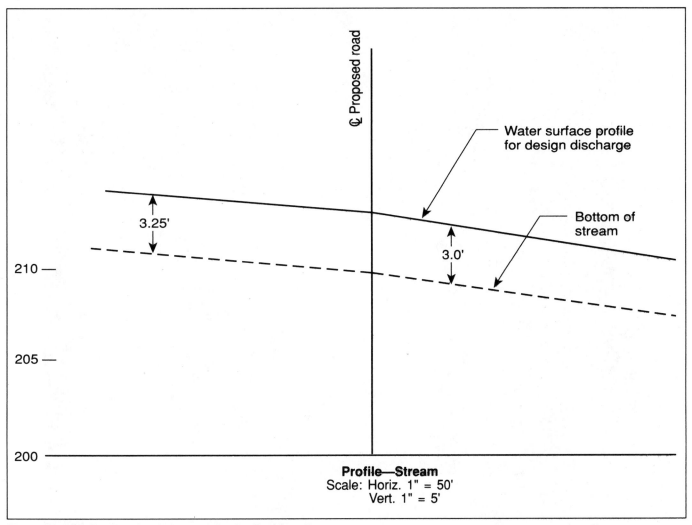

Figure 6-30 Profile of existing stream for Problem 5.

FURTHER READING

U.S. Department of Transportation, Federal Highway Admin. (1967). *Use of Riprap for Bank Protection*, Hydraulic Engineering Circular No. 11. Washington, D.C.: Dept. of Transportation.

U.S. Department of Transportation, Federal Highway Admin. (1972). *Hydraulic Design of Improved Inlets for Culverts*, Hydraulic Engineering Circular No. 13. Washington, D.C.: Dept. of Transportation.

U.S. Department of Transportation, Federal Highway Admin. (1975). *Design of Stable Channels with Flexible Linings*, Hydraulic Engineering Circular No. 15. Washington, D.C.: Dept. of Transportation.

CHAPTER

7

DETENTION DESIGN

Stormwater management is a term used to describe all endeavors to control runoff in areas affected by development. Typical measures include storm sewers, culverts, swales and detention basins. Storm sewers and culverts help convey stormwater from one point to another, but detention basins actually reduce the peak rate of flow. In addition, detention basins play a crucial role in reducing the amount of pollutants suspended in stormwater. So, in short, detention basins are structures that control both the quantity and the quality of stormwater runoff and therefore have become the central focus of stormwater management.

In this chapter, we will learn the fundamental principles governing the functioning of a detention basin and its application to stormwater management. Basic principles of design of detention basins will be developed through examples involving subdivisions and site plans. As in previous chapters, design considerations are limited mostly to hydraulic design and do not emphasize other related design factors such as structural stability and cost.

OBJECTIVES

After completing this chapter, the reader should be able to:
- Explain the difference between on-site and regional detention
- Compute impoundment volume by the elevation-area method
- Compute impoundment outflow using orifice and weirs
- Compute a reservoir routing by hand
- Design a basic detention basin
- Relate detention basin design to actual case studies

7.1 FUNDAMENTAL CONCEPTS

The primary task of a detention basin is to temporarily store water during a storm and then release it slowly so that the net effect is a reduction of the rate of flow of the runoff. This reduction is called an *attenuation* of the flow. Water is stored in a pond or reservoir, which is usually an open cut in the ground or a series of underground pipes or chambers. The slow release of water is accomplished by an outlet structure or spillway employing an orifice or weir or a combination of flow controlling devices. Figure 7-1 shows schematic sketches of typical detention basin arrangements.

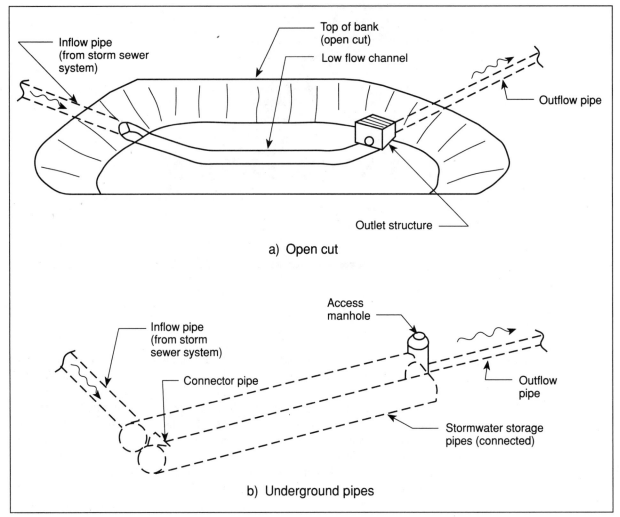

Figure 7-1 Typical detention basin facilities.

Detention facilities usually are required by local ordinance to be part of the stormwater management measures for all land development. For the purpose of our analysis, we will consider two types of development: residential subdivision and commercial site plans.

Most detention basins are designed as an open cut in the ground with or without a pond at the bottom. The open cut provides the volume required to temporarily store stormwater during a storm. However, when space on a site is not available, some detention basins are formed by placing large diameter pipes or other types of chambers underground. The volume provided by the pipes or chambers takes the place of the open cut basin. These types of basins are less desirable because generally pipes provide much less volume than does an open cut and therefore the cost can skyrocket. Furthermore, underground basins do not possess the pollution trapping ability of an open cut basin, thus requiring additional measures to control pollution. This causes another increase in cost.

On-Site Detention

Detention basins can be designed to control the runoff from a single site or tract of land or from an entire region of a municipality or county. In the latter case, the basin is called a regional detention basin. In Figure 7-2, we see a small on-site basin that has become a familiar sight in commercial as well as residential developments.

Figure 7-2 Typical small on-site detention basin. A riprap low-flow channel runs down the center of the basin.

When a detention basin is designed for a single site, the usual design philosophy is:

1. The peak rate of stormwater leaving the site shall not be greater after development than before.
2. Stormwater leaving the site shall not adversely affect adjacent properties.
3. Soil erosion during and after development shall not increase over that which naturally occurs.
4. Natural drainage patterns shall not be significantly altered.

Let us take a closer look at this philosophy and its implications, starting with Item 1. The natural drainage pattern of a site may be such that runoff leaves the site at only one point, or it could be that runoff exits the site at two or more locations. In either case, it is the total peak runoff for all points of exit for existing and proposed conditions that is compared. In Case Study 1 (Section 7.6) we will analyze a site that drains in two directions and see how the designer copes with this situation.

Item 2 above asserts that stormwater leaving the site should not adversely affect adjacent properties. This is important because of the way storm sewers can concentrate the flow of stormwater. So even if the total peak flow of the site is reduced, the peak flow at any particular point of exit could increase and possibly cause erosion or flooding.

Item 3 above states that soil erosion should not be allowed to increase. This is accomplished by employing special erosion control measures during construction and providing proper lining after construction. Protective lining can consist of grass, concrete or riprap in addition to other materials.

According to Item 4, drainage patterns should not be significantly altered. This means that if one portion of the site is tributary to one river and the remainder is tributary to another river, proposed grading on the site should not shift the divide separating the two watersheds, thus putting more runoff into one river and taking it away from the other. However, minor shifting of the watershed divide is allowed and, in fact, usually is unavoidable.

Regional Detention

As stated earlier, most detention basins are constructed on individual sites and designed to control the runoff from that site alone. However, calculations performed on a regionwide basis show that individual sites are not always the best place to control the quantity of runoff. In assessing the optimum location for detention facilities, we must focus on the impact point where we wish to attenuate the stormwater flow. Is it immediately downstream of the site or is it a mile downstream?

Consider the industrial subdivision shown in Figure 7-3. Stormwater outflow from the subdivision is piped directly to the Black River without affecting any adjacent property. The off-site pipe discharging to the river is designed to convey the 100-year storm flow. Should a detention basin be constructed on-site in accordance with local ordinance? To answer this question, consider the off-site impact. The Black River has capacity to carry the 100-year flood in the project vicinity, but 0.75 miles downstream low banks result in a flooding condition. The flow contribution of the subdivision to the Black River at the point of flooding is plotted on the hydrograph in Figure 7-4. If no detention basin is constructed, runoff from the site will peak early and its contribution at the impact point will be relatively low. But if a detention basin is constructed, the outflow will peak in the Black River at a later time and even though the attenuated flow has a lesser peak contribution than flow with no detention, its contribution at time t_1 when the river flow peaks is greater. Thus, the presence of a detention basin in this case actually aggravates the flooding condition at the most vulnerable point downstream. For this reason, the industrial subdivision should not include a detention basin for stormwater quantity control.

Other land development locations will not present the same downstream conditions as the industrial tract just described. Sometimes, on-site detention makes the most sense. Sometimes regional detention is preferable. As shown in Figure 7-5, for example, a watershed tributary to the Blue River could be controlled by twelve on-site detention basins. But

Figure 7-3   Plan of industrial development with runoff discharging to an adjacent river. What purpose does an on-site detention basin serve in this case?

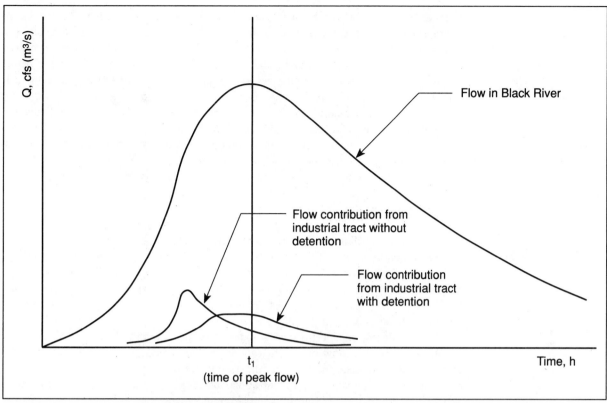

Figure 7-4   Hydrograph of flood flow in Black River at point of flooding as shown in Figure 7-5.

Figure 7-5   On-site detention basins versus regional facilities.

detention basins 1 through 6 could be replaced by one regional basin R1 and basins 7 through 11 could be replaced by regional basin R2. Or all twelve basins could be replaced by regional basin R3. This consolidation of detention facilities could be less expensive, require less space and be easier to maintain than individual on-site basins. Of course, before initiating a regional stormwater management system like that in Figure 7-5, all vulnerable impact locations would have to be analyzed. It could be that local flooding within the watershed requires numerous small detention basins. Also, perhaps basins R1 and R3, which are on-stream, will have an adverse effect on fish habitat.

Regional detention is difficult to accomplish but often is more effective and less costly than allowing a proliferation of small on-site facilities. However, the designer has little choice in most cases and must follow local regulations.

Water Quality Control

In addition to controlling the quantity of runoff, detention basins can serve effectively to control the quality as well. When stormwater runs across man-made impervious surfaces, it picks up pollutants which become suspended in the flow stream. The pollutants typically consist of hydrocarbons, phosphates, nitrates, salts, pesticides and heavy metals. The mechanism of transport for most pollutants is to adhere to silt particles as the silt is carried along suspended in the stormwater.

Most pollutant transport occurs at the early stage of a rainfall event. Following the initial losses, the first flow of runoff, called *first flush*, picks up the loose dust that has coated the ground since the last rainfall. As the storm progresses and heavier flows of runoff occur, the amount of pollutant transport lessens, making this later runoff much cleaner than the first flush. Since this initial pollutant carrying discharge is relatively low in flow rate, you can imagine that it does not take a large storm to transport most of the pollutants. It is for this reason, therefore, that the smaller storms such as the 1-year storm are typically controlled in an effort to manage pollution.

A detention basin controls pollutants by trapping the silt upon which the pollutants ride. As the first flush flow enters into a grass-lined detention basin, the suspended silt is filtered out by the grass; then as the water starts to collect or form a pool, silt particles settle out as they travel through the pool toward the outlet. Trapped pollutants then follow one of these fates: they infiltrate into the ground and lodge in the void spaces between soil particles or they are absorbed by the grass roots and then harvested with the grass or, if they are volatile, they evaporate.

Not all pollutants are trapped by the typical detention basin. If the basin is constructed to have a permanent pond at its bottom, trap efficiency is increased because the settling process is enhanced. This type of detention basin is called a *wet basin, retention basin* or *sediment basin*. However, studies show that for typical wet basins without extraordinary contrivances, pollutant reduction does not exceed 60%. Figure 7-6 shows a schematic sketch of a wet basin.

Figure 7-6   Cross section of typical wet detention basin.

Infiltration Basin

Another type of detention basin, called an *infiltration basin*, adds a third function to those already discussed. In addition to controlling the quantity and quality of stormwater, it promotes recharge of stormwater to groundwater storage. The advantage of recharge is that it helps prevent the depletion of groundwater storage. This is very important to drinking water supply.

One of the consequences of land development not yet mentioned is that a greater overall volume of water is converted to runoff and less water infiltrates the ground to become groundwater. Referring to Figure 4-2, we see that precipitation has three fates:

1. overland flow (runoff),
2. subsurface flow, and
3. groundwater.

Land development, due to the addition of impervious surface, upsets this distribution by increasing the percentage of runoff and lessening subsurface flow and groundwater recharge.

So to help re-establish the natural distribution of precipitation, recharge basins employ special measures to intercept some of the runoff and place it into the groundwater. All open cut detention basins accomplish some groundwater recharge, but the percentage usually is negligible unless special measures are employed. These measures typically consist of a depression in the ground below the outlet elevation, much like a retention basin, together with excavated porous connections to the ground so that the water retained in the pool infiltrates into it through the excavation. Figure 7-7 shows a schematic diagram of a typical recharge basin.

Figure 7-7 Cross section of typical recharge basin.

Figure 7-8 Cross-section views of two examples of open-cut impoundments.

7.2 STORMWATER IMPOUNDMENT

The *impoundment* is the volume of water temporarily stored in a detention basin during a rainfall event. In the case of a wet basin, it is the volume of water stored above the permanent water level; that is, not including the permanent pool. The usual impoundment is provided by excavating an open cut in the ground and allowing stormwater to accumulate in the open cut. In some cases, an earth berm must be constructed at one end of the detention basin to provide sufficient storage volume. Figure 7-8 shows two cases of open cut impoundments, one requiring an earth berm and one not needing a berm.

The usual method for computing storage volume is the elevation-area method. In this method, contour lines are traced around the proposed basin and the area contained within each contour measured. A table of elevations and corresponding areas is then made. The volume contained between any two adjacent contours is estimated as the average of the two areas multiplied by the vertical distance between them. Each volume thus computed is an incremental volume and the total volume is the sum of the incremental volumes.

The elevation-area method is illustrated in the following example.

Example 7-1

Problem

Find the volume contained in the open-cut detention basin shown in Figure 7-9 up to the elevation 236. Areas contained within the contours were measured by planimeter as follows:

Elevation (ft)	Area (ft²)
230	0
231	250
232	840
233	1350
234	2280
235	3680
236	5040

Solution

Prepare a chart with four columns as shown below, the first two columns being a copy of the data shown above.

(1) Elevation (ft)	(2) Area (ft²)	(3) Incremental Volume (ft³)	(4) Cumulative Volume (ft³)
230	0	0	0
231	250	125	125
232	840	545	670
233	1350	1095	1765
234	2280	1815	3580
235	3680	2980	6560
236	5040	4360	10920

Each value in Column 3 represents the volume contained between that elevation and the next lower elevation. The first value is zero because elevation 230 is the lowest elevation and there is no volume contained beneath it. The second value in Column 3 is found by averaging the first two values in Column 2 and multiplying by the difference between the first two elevations in Column 1:

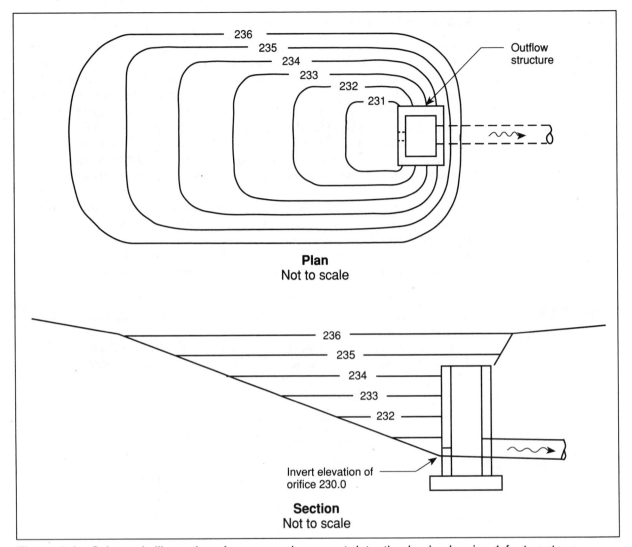

Figure 7-9 Schematic illustration of a proposed open-cut detention basin showing 1-foot contour intervals.

$$\frac{250 + 0}{2} \times 1 = 125 \text{ ft}^3.$$

The third value in Column 3 is computed by averaging the second and third values in Column 2 and multiplying by the difference between the second and third elevations in Column 1:

$$\frac{840 + 250}{2} \times 1 = 545 \text{ ft}^3.$$

The rest of the values in Column 3 are computed in like manner.

Each value in Column 4 represents the sum of all values in Column 3 up to and including that elevation. The first value is zero because the first value in Column 2 is zero. The second value is computed by adding the first two values in Column 3:

0 + 125 = 125 ft³.

The third value is computed by adding the first three values in Column 3:

0 + 125 + 545 = 670 ft³.

The remaining values in Column 4 are computed in like manner.

The total volume up to elevation 236 is the value in Column 4 opposite elevation 236, or 10,920 ft³. (Answer)

For underground detention basins using pipes for storage, the elevation-area method for computing volume is not convenient, so another method must be used. The easiest computation is by the average end-area method which is illustrated by the following example.

Example 7-2

Problem
Find the volume contained in the pipe detention basin shown in Figure 7-10.

Solution
Prepare a chart with four columns as shown below.

(1) Elevation (ft)	(2) Downstream Area (ft²)	(3) Upstream Area (ft²)	(4) Volume (ft³)
520	0	0	0
521	2.39	0	239
522	6.29	2.39	868
523	10.1	6.29	1639
524	12.6	10.1	2270
525	12.6	12.6	2520

Each value in Column 2 represents the area of the segment of a circle related to the depth of water at the downstream end of the pipe. (The pipe is designed with a slight gradient so that it will completely drain after each storm.)

The first value is zero because elevation 520 is at the invert of the pipe. To find the second value in Column 2, use Figure 3-9. Elevation 521 represents a depth of 1.0 foot which is 25% of the pipe diameter. Therefore, enter Figure 3-9 at 25% depth of flow and extend a line to the right until it intersects the area line. From this intersection, drop a line straight down and read the percentage on the Hydraulic Elements scale. The resulting area is 19% of the full cross-sectional area, which is 12.57 ft². Therefore the area is

(12.57) (.19) = 2.39 ft².

The third value in Column 2 corresponds to elevation 522, which represents a depth of 2.0 feet, 50% of the pipe diameter. Therefore, enter Figure 3-9 at 50% and read 50% on the Hydraulic Elements scale. The area is thus

(12.57) (.50) = 6.29 ft².

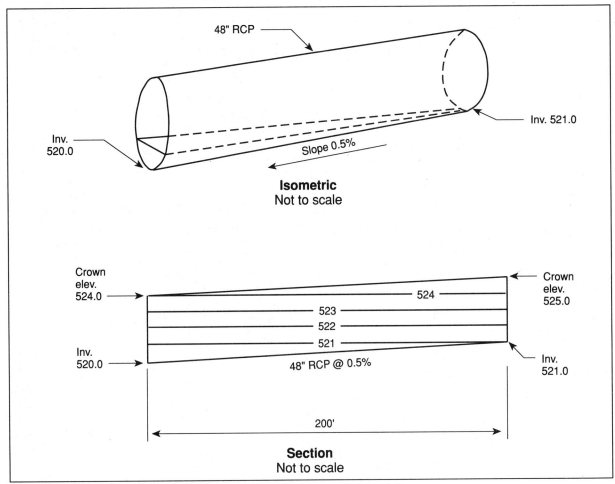

Figure 7-10 Schematic illustration of a proposed pipe detention basin.

The fourth value corresponds to elevation 523, which represents a depth of 3.0 feet, 75% of the pipe diameter. Entering Figure 3-9 at 75%, the Hydraulic Elements scale reads 80%. Therefore, the area is

$$(12.57) \ (.80) = 10.1 \ \text{ft}^2.$$

The fifth value corresponds to elevation 524, which represents full depth at the downstream end of the pipe. Therefore, the area is 12.6 ft².

The sixth value corresponds to elevation 525 which also represents full depth at the downstream end of the pipe. Therefore, the area again is 12.6 ft².

Each value in Column 3 represents the area of the segment of a circle related to the depth of water at the upstream end of the pipe. The first value is zero because there is no water in the pipe. The second value also is zero because when the water level reaches elevation 521, it just reaches the invert at the upstream end which represents a depth of zero foot at that end.

The third value in Column 3 corresponds to elevation 522 which represents a depth of 1.0 foot. Using Figure 3-9, the area is 19% of full area, or 2.39 ft².

The remaining values in Column 3 are determined in the same manner as Column 2.

Each value in Column 4 is found by computing the average of the two corresponding end areas in Columns 2 and 3 and then multiplying by the distance between the ends. The first value is

$$\frac{0+0}{2} \times 200 = 0 \text{ ft}^3.$$

The second value is

$$\frac{2.39+0}{2} \times 200 = 239 \text{ ft}^3.$$

The third value is

$$\frac{6.29+2.39}{2} \times 200 = 868 \text{ ft}^3.$$

The remaining values are determined using the same method. The total volume is the value in Column 4 corresponding to the highest elevation of the pipe, or 2,520 ft³. (Answer)

> *Note:* It would be much faster to compute total volume as you would a cylinder, namely end area multiplied by length. However, the full tabular format is needed for subsequent detention basin design calculations.

7.3 OUTLET STRUCTURE

Impounded water in a detention basin is released slowly through the outlet structure, which usually consists of an orifice or combination orifice and weir. The simplest outlet structure consists of a single pipe with invert set at the lowest elevation of the basin. This is called a single-stage outlet because water is allowed to flow out through only one opening. However, another higher outlet generally is provided as an emergency relief to be used only if the outlet structure cannot handle the entire storm runoff. Emergency spillways are analyzed separately in the next section.

Figure 7-11 shows a single-stage outlet in the form of a single pipe installed in the berm of a detention basin. Although an open cut basin is depicted, single stage outlets can be used with underground detention as well.

The sketch in Figure 7-11 suggests that this single stage outlet resembles a culvert, and indeed, it does act hydraulically as a culvert, usually with inlet control. However,

Figure 7-11   Typical single stage outlet.

such a single stage arrangement is usually modeled as an orifice because the orifice model yields discharges very similar to a culvert acting with inlet control, especially for impoundment depths above the pipe crown.

As part of the detention basin calculations, a *discharge rating* of the outlet is computed. That is, for a series of hypothetical water levels in the impoundment, corresponding discharge through the orifice is determined. The following example illustrates this computation.

Example 7-3

Problem

An outlet consisting of a 12-inch pipe is proposed for the detention basin shown in Figure 7-11. The invert of the pipe is 322.0 and the top of berm is set at elevation 327.0. Compute the discharge rating for the outlet.

Solution

Assume that the pipe opening is a 12-inch orifice and that discharge is computed using Equation 2-13. Then compute discharge for water level elevations of 322, 323, 324, 325, 326 and 327.

For water level elevation 322, head is equal to zero because water level is at the invert of the orifice. Discharge therefore also is zero.

For water level elevation 323, head is determined to be 0.5 feet as shown below.

The area of the orifice is

$$a = \frac{\pi(1.0)^2}{4} = 0.785 \text{ ft}^2$$

and the discharge, Q, is

$$Q = ca\sqrt{2gh}$$

$$Q = (.62)(.785)\sqrt{2(32.2)(.5)}$$

$$Q = 2.76 \text{ cfs.} \quad \text{(Answer)}$$

For water level elevation 324, head is determined to be 1.5 feet as shown below.

Therefore, discharge is

$$Q = ca\sqrt{2gh}$$

$$Q = (.62)\,(.785)\,\sqrt{2(32.2)\,(1.5)}$$

$$Q = 4.78 \text{ cfs.}\quad \text{(Answer)}$$

For water level elevation 325, head is determined to be 2.5 feet. Therefore, discharge is

$$Q = (.62)\,(.785)\,\sqrt{2(32.2)\,(2.5)}$$

$$Q = 6.18 \text{ cfs.}\quad \text{(Answer)}$$

For water level elevation 326, head is determined to be 3.5 feet. Therefore, discharge is

$$Q = (.62)\,(.785)\,\sqrt{2(32.2)\,(3.5)}$$

$$Q = 7.31 \text{ cfs.}\quad \text{(Answer)}$$

Finally, for water level elevation 327, head is determined to be 4.5 feet and discharge is computed as

$$Q = (.62)\,(.785)\,\sqrt{2(32.2)\,(4.5)}$$

$$Q = 8.29 \text{ cfs.}\quad \text{(Answer)}$$

These results are then summarized in the following table:

W.L. Elev. (ft)	12" Orifice	
	h (ft)	Q (cfs)
322	0	0
323	0.5	2.76
324	1.5	4.78
325	2.5	6.18
326	3.5	7.31
327	4.5	8.29

Multi-Stage Outlet

Often, a detention basin design requires a two-stage outlet, or multiple stages beyond two, to create a special distribution of discharges to meet particular design conditions. Figures 7-12 and 7-13 show typical two-stage and three-stage outlet structures, respectively.

To compute a discharge rating for this type of outlet structure, each stage is considered separately and then all discharges are added to give the total discharge.

Figure 7-12   Typical two-stage outlet structure.

Figure 7-13 Typical three-stage outlet structure.

The following example illustrates typical computations for a multiple stage outlet structure.

Example 7-4

Problem

A three-stage outlet structure similar to that shown in Figure 7-13 is proposed for a detention basin design. Key elevations are listed below.

1. Primary stage (4-inch orifice)—Invert elevation 560.0
2. Second stage (1.5-foot weir)—Crest elevation 562.67
3. Third stage (12.5-foot weir)—Crest elevation 563.67
4. Top of berm—elevation 565.0
5. Outflow pipe—(48-inch diameter)—Invert elevation 558.0

Note: For this example, assume no emergency spillway.

Compute the discharge rating for the outlet structure.

Solution

To create a discharge rating, compute discharge for a series of elevations ranging from the bottom of the basin (elevation 560.0) to the top of the berm (elevation 565.0). The elevations can be every whole foot but should include each stage along the way. Thus, the water level elevations to be considered are 560, 561, 562, 562.67, 563.67, 564 and 565.

For water level elevation 560, head for the orifice is equal to zero because water level is at the invert of the orifice. Discharge therefore also is zero.

For water level elevation 561, head is determined to be 0.83 feet as shown below.

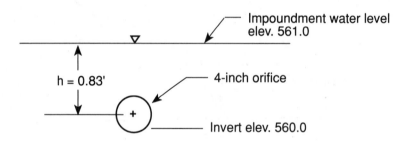

The area of the orifice is

$$a = \frac{\pi(.333)^2}{4}$$

$$a = 0.087 \text{ ft}^2,$$

and the discharge, Q, is

$$Q = ca\sqrt{2gh}$$

$$Q = (.62)(.087)\sqrt{2(32.2)(.83)}$$

$$Q = 0.39 \text{ cfs.}$$

For water level elevation 562, head is determined to be 1.83 feet because it is 1.0 foot higher than the previous value of head. Therefore, discharge is

$$Q = (.62)(.087)\sqrt{2(32.2)(1.83)}$$

$$Q = 0.59 \text{ cfs.}$$

For water level elevation 562.67, head is determined to be 2.5 feet. Therefore discharge is

$$Q = (.62)(.087)\sqrt{2(32.2)(2.5)}$$

$$Q = 0.68 \text{ cfs.}$$

For water level elevation 563.67, head is determined to be 3.5 feet. Therefore, discharge is

$$Q = (.62)(.087)\sqrt{2(32.2)(3.5)}$$

$$Q = 0.81 \text{ cfs.}$$

For water level elevation 564, head is determined to be 3.83 feet. Therefore, discharge is

$$Q = (.62)(.087)\sqrt{2(32.2)(3.83)}$$

$$Q = 0.85 \text{ cfs.}$$

For water level elevation 565, head is determined to be 4.83 feet. Therefore, discharge is

$$Q = (.62)(.087)\sqrt{2(32.2)(4.83)}$$

$$Q = 0.95 \text{ cfs.}$$

Next, discharge over the second stage weir is computed using Equation 2-14. For water level elevations 560, 561 and 562, discharge is zero because the water is below the weir crest. At elevation 562.67, water level is just at the crest, so head is equal to zero, and once again discharge is zero.

For water level elevation 563.67, head is determined to be 1.0 foot. Since the outlet structure wall thickness is 0.5 foot, discharge coefficient, c, is found in Appendix A-5 to be 3.32 for all values of head equal to and greater than 1.0 foot. Therefore, discharge is

$$Q = cLH^{3/2}$$

$$Q = (3.32)(1.5)(1.0)^{3/2}$$

$$Q = 4.98 \text{ cfs.}$$

For water level elevation 564, head is determined to be 1.33 feet. Therefore, discharge is

$$Q = (3.32) \ (1.5) \ (1.33)^{3/2}$$

$$Q = 7.64 \ \text{cfs.}$$

For water level elevation 565, head is determined to be 2.33 feet. Therefore, discharge is

$$Q = (3.32) \ (1.5) \ (2.33)^{3/2}$$

$$Q = 17.7 \ \text{cfs.}$$

Next, discharge over the third stage weir is computed. Discharge is zero for water level elevations 560, 561, 562, 562.67 and 563.67.

For water level elevation 564, head is determined to be 0.33 feet. Therefore, discharge is

$$Q = (3.32) \ (12.5) \ (0.33)^{3/2}$$

$$Q = 7.87 \ \text{cfs.}$$

For water level elevation 565, head is determined to be 1.33 feet. Therefore, discharge is

$$Q = (3.32) \ (12.5) \ (1.33)^{3/2}$$

$$Q = 63.7 \ \text{cfs.}$$

These results are then summarized in the following table:

W.L. Elev (ft)	4" Orifice		1.5' Weir		12.5' Weir		Total Q (cfs)
	h (ft)	Q (cfs)	H (ft)	Q (cfs)	H (ft)	Q (cfs)	
560	0.0	0.0	—	—	—	—	0.0
561	.83	.39	—	—	—	—	.39
562	1.83	.59	—	—	—	—	.59
562.67	2.5	.68	0.0	0.0	—	—	.68
563.67	3.5	.81	1.0	4.98	0.0	0.0	5.79
564	3.83	.85	1.33	7.64	.33	7.87	16.36
565	4.83	.95	2.33	17.7	1.33	63.7	82.4

Finally, check the outflow pipe to be sure it has the capacity to handle the discharge coming into the outflow structure. Treating the outflow pipe as a culvert operating with inlet control, headwater for the maximum Q of 82.4 cfs is found from Chart 2 of

Appendix B-1 to be 4.44 feet. This gives a water level elevation inside the outlet structure of

$$558.0 + 4.44 = 562.44$$

as shown below.

If the water level inside the outlet structure rises above the crest of the second stage weir, the weir would be submerged and its discharge reduced. This would render the discharge rating somewhat invalid at the highest elevations. Therefore, the outlet pipe should be designed to adequately handle the maximum discharge without causing an excessive back-up into the outlet structure.

The 48-inch RCP outlet pipe positioned with its invert at 558.0 is adequate.

However, it should be noted that when the water level rises in the outlet structure, discharge through the orifice is altered. When the orifice becomes submerged, head is measured to the downstream water level, not the center of the orifice, thus reducing h and consequently reducing Q.

Thus, values of Q for the orifice at the higher water levels were computed too high. However, in this case, such an error is not significant since the reduction of Q is less than 1 cfs when total Q is in the range of 30 cfs and more.

7.4 EMERGENCY SPILLWAY

As a safety feature to prevent detention basins from overflowing, an *emergency spillway* is included in most designs. Emergency spillways consist of an additional outlet set at an elevation higher than all other outlets so that water will not enter unless the impoundment level has risen higher than anticipated for the design storm.

Impoundment level could exceed the design maximum elevation for a variety of reasons; principal among them being the following:

1. A rainfall could occur which exceeds the design storm rainfall.

2. The outlet structure could become blocked, thus preventing stored water from exiting the basin fast enough.

The primary purpose of an emergency spillway is to provide a way for excessive water in the impoundment to safely exit, thus preventing an overtopping of the berm. If the berm is overtopped it could be breached, or washed out, allowing impounded water to suddenly flood areas downstream of the basin. If a detention basin has no berm and an overtopping of the banks would cause no breach or failure of the sides of the basin then the need for an emergency spillway is lessened.

A secondary purpose of an emergency spillway is to control any eventual overtopping of the detention basin by directing the flow in a harmless direction thus preventing uncontrolled flow even if no breach occurs.

Most detention basin designs incorporate one of two general types of emergency spillway. These are

1. Grassed channel or waterway located separate from the outflow structure.
2. Allowing water to flow into the top of the outflow structure and through the outflow pipe.

Separate grassed channels are used for major detention basins with large maximum impoundments because this is the safest type of emergency spillway. Even if the outflow pipe becomes totally clogged, the grassed channel is free to function and safely convey the excess flow out of the basin. The channel must have an entrance invert below the top of berm and it must be located in virgin ground beyond the berm. Normally, the channel cannot be formed by excavating a cut in the berm. This would be feasible only if special measures are employed to protect the berm against erosion.

Figure 7-14 shows a typical emergency spillway arrangement using a grassed swale.

Figure 7-14 Typical detention basin with grassed channel emergency spillway.

Design criteria for an emergency spillway vary considerably but can be summarized as follows:

1. The invert or crest of the emergency spillway is set at or above the maximum impoundment elevation computed for the design storm.

2. a) The emergency spillway is designed to convey a storm with greater rainfall than the design storm. This larger storm is usually called the *emergency spillway design storm*. The emergency spillway would convey whatever is not handled by the outflow structure.

or

   b) The emergency spillway is designed to convey the peak discharge of the design storm. This would simulate a situation in which all flow exits the basin through the emergency spillway.

3. The top of berm elevation is set at the elevation of maximum flow through the emergency spillway plus an additional vertical distance called *freeboard*.

Freeboard provides an extra measure of safety to account for all contingencies such as a storm greater than the design storm or errors in calculations or errors in construction. Freeboard amount can range from zero to one foot or more.

Figure 7-15(a) shows a typical detention basin outlet with separate emergency spillway. Even though the emergency spillway is shown in the same cross section as the outlet structure, it is actually at a different location along the berm. Notice that the crest of the emergency spillway is set at the same elevation as the maximum design water level. This is the impoundment level that will result if the design storm occurs and the basin functions properly.

Figure 7-15 Typical emergency spillway arrangements.

As soon as the design maximum impoundment level is exceeded, water starts entering the emergency spillway. So the elevation of the emergency spillway cannot be determined until the basin calculations have determined the maximum impoundment level. Likewise, the top of berm elevation cannot be determined until the water level in the emergency spillway is computed. Detention basin calculations are presented in the next section.

As mentioned earlier, the emergency spillway can be incorporated into the outlet structure in certain cases. Figure 7-15(b) depicts this type of arrangement. Notice that the top of the outlet structure is set at the same elevation as the maximum design water level. This is because the opening at the top of the structure is being used as the emergency spillway. When the impoundment rises above this level, water starts to pour into the top of the structure which acts as emergency spillway.

In this type of arrangement, the outflow pipe must handle all outflow including normal outflow as well as emergency outflow. Therefore, as you might expect, the pipe must be designed larger than in the case of a separate emergency spillway.

In Example 7-4, since the outflow structure handled impoundment levels up to the top of berm, the outlet pipe was as large as it would have to be if the emergency spillway was incorporated into the outlet structure.

7.5 RESERVOIR ROUTING

When runoff enters a detention basin and is temporarily stored and then exits through the outlet structure, the peak rate of outflow is less than the peak rate of inflow. This reduction in peak flow is called attenuation and the procedure for computing the outflow hydrograph when the inflow hydrograph is known is called *routing*. The term routing describes a mathematical procedure, not the mapping of a course of movement.

The term attenuation was first discussed in Section 4.7 in connection with stormwater flowing in a stream. Whether stormwater flows along a stream or through a detention basin, its hydrograph is attenuated; that is, the hydrograph is lowered and spread out as shown in Figure 4-23.

Several methods have been devised over the years to compute stream and reservoir (detention basin) routing. For stream routings, the Muskingum Method and the Kinematic Wave Method are widely used. For reservoir routing, the Modified Puls Method is commonly employed.

This routing method relies on the so-called continuity equation which is a statement of conservation of the mass of water entering and leaving the reservoir or impoundment. (The continuity equation should not be confused with Q = va which goes by the same name in a different context.)

The continuity equation is expressed as

$$\bar{I} - \bar{O} = \frac{\Delta S}{\Delta t} \tag{7-1}$$

where $\bar{I}$ = mean flow into reservoir during time Δt, cfs (m³/s)
  $\bar{O}$ = mean outflow from reservoir during time Δt, cfs (m³/s)
  ΔS = change in reservoir storage during time Δt, ft³ (m³)
  Δt = incremental time period, s.

The equation can be considered to be a ledger that keeps account of the balance between water entering, water leaving and water stored much like the constantly changing inventory of a warehouse where goods come in and go out and are stored. Numerical solution of the continuity equation is an iteration process in which a small time increment is chosen and the volume balance computed at the end of each time period.

So, to use the analogy of a warehouse with a time increment of 1 day, if during the first day 1000 boxes come in and 400 boxes go out, the change in storage is positive 600 boxes. If, during the second day, 500 boxes come in and 700 boxes go out, the change in storage is negative 200 boxes and the remaining inventory is 400 boxes.

To solve Equation 7-1, first re-write the equation in a more useful form

$$\frac{I_1 + I_2}{2} - \frac{O_1 + O_2}{2} = \frac{S_2 - S_1}{\Delta t} \tag{7-2}$$

where the subscripts 1 and 2 denote the beginning and end, respectively, of the chosen time period Δt. Terms in Equation 7-2 may now be rearranged as

$$(I_1 + I_2) + \left[\frac{2S_1}{\Delta t} - O_1\right] = \frac{2S_2}{\Delta t} + O_2. \tag{7-3}$$

In Equation 7-3, all terms on the left-hand side are known from preceding routing computations while the right-hand terms are unknown and must be determined by storage routing.

Assumptions implicit in this routing method are:

1. The reservoir water surface is horizontal.
2. The outflow is a unique function of storage volume.
3. Outflow rate varies linearly with time during each time period Δt.

In detention basin design, Equation 7-3 is used to compute the outflow hydrograph when the inflow hydrograph is known. This computation constitutes a routing. If the specified detention basin does not produce the desired results, the parameters must be revised and another routing performed. Thus, the process is one of trial and error, just as with culvert design.

Now, let's take a look at Equation 7-3 to see how such an equation can be solved. The first question is: What result do we want the equation to give us? The answer is: an outflow hydrograph, that is, a complete list of outflow rates from the detention basin during the design storm. These values are produced by the term O_2 in Equation 7-3. Remember that the equation will be solved many times, once for every chosen time period throughout the storm. Each solution of the equation gives another value O_2 and the total list of such values comprises the outflow hydrograph.

But both S_2 and O_2 are unknowns each time Equation 7-3 is solved. To overcome this problem, we take advantage of the fact that S_2 and O_2 are related to each other, which in effect supplies a second equation which is needed to solve for two unknowns.

Now, again looking at Equation 7-3, the first term $(I_1 + I_2)$ comes from the inflow hydrograph, which is the runoff hydrograph for the design storm for the watershed tributary to the detention basin. The parameter Δt as mentioned is an arbitrary time period, which should be chosen as small as practical remembering that the smaller Δt, the more time periods and the more computations. Δt should be chosen small enough to create at least 4 or 5 points on the rising limb of the inflow hydrograph with one point coinciding with peak inflow.

The parameter S_1 represents the storage volume in the detention basin at the beginning of each time period. It is determined simply by taking the ending storage volume from the preceding time period.

Finally, the parameter O_1 represents the outflow at the beginning of each time period and is determined simply by taking the outflow at the end of the time period preceding.

Row	(1) Time (hr)	(2) I_1 (cfs)	(3) $I_1 + I_2$ (cfs)	(4) $\frac{2S}{\Delta t} - O$ (cfs)	(5) $\frac{2S}{\Delta t} + O$ (cfs)	(6) O_2 (cfs)
1						
2						
3						

Figure 7-16   Table headings for hand reservoir routing.

To carry out the iterative solution of Equation 7-3 therefore you must start with three elements:

1. Inflow hydrograph
2. Relation of storage volume to elevation in the proposed detention basin
3. Relation of outflow to water level elevation in the proposed detention basin (outflow rating)

Based on the duration of the rising limb of the inflow hydrograph, choose a time period, Δt. Then, combine the information in Items 2 and 3 above to create two new relationships, namely graphs of O vs $2S/\Delta t + O$ and O vs $2S/\Delta t - O$.

Next, set up a table with headings like those in Figure 7-16.

In Column 1, all time values, generated as multiples of Δt, are listed. In Column 2, values of the inflow hydrograph corresponding to the time values in Column 1 are listed. In Column 3, successive pairs of inflow values are summed and listed; that is, in Row 1, the sum of the two I_1 values in Rows 1 and 2 is listed, and so on. In Columns 4 and 5, values of $2S/\Delta t - O$ and $2S/\Delta t + O$ are generated from the appropriate graph or from Equation 7-3. In Column 6, values of O_2 are generated, that is, the value of O_2 in Row 2 represents the outflow at the end of the time period starting with Row 1 and ending at Row 2. The use of this routing table is demonstrated in the following example.

Example 7-5

Problem

A detention basin is specified in Figure 7-17 and an inflow hydrograph in Table 7-1. Route the hydrograph through the basin to produce the resulting outflow hydrograph.

Table 7-1
Inflow hydrograph for Example 7-5

Time (hr)	Inflow (cfs)
0.0	4
.4	6
.8	9
1.2	23
1.6	84
2.0	48
2.4	20
2.8	14
3.2	11
3.6	9
4.0	8
4.4	7
4.8	7

PLAN
N.T.S.

STORAGE VOLUME			DISCHARGE RATING	
Elevation (ft)	volume (ft³)		Elevation (ft)	Outflow (cfs)
100	0		100	0
102	7500		102	2.5
104	36000		104	24.9
105	60000		105	67.5
106	90000		106	122

Figure 7-17   Detention basin parameters for Example 7-5.

Solution

As shown in Table 7-1, the incremental time period was chosen as $\Delta t = 0.40$ hour, which is equal to 1440 seconds. Now, using the information in Figure 7-17, create a table of values of outflow, O versus $2S/\Delta t - O$ and versus $2S/\Delta t + O$. This is depicted in Table 7-2.

Next, sketch a graph of the above values. The graph is shown in Figure 7-18. Now create a table like the one shown in Figure 7-16 with the inflow hydrograph listed in Columns 1 and 2. The table is depicted in Table 7-3.

Fill in the appropriate numbers for Column 3 by simply adding successive pairs of numbers in Column 2. For example, the value in Column 3, Row 5 is the sum of the numbers in Column 2, Rows 5 and 6.

Now, the remaining 3 columns must be generated one row at a time. To begin, consider the first time period, which starts at Row 1 and ends at Row 2. The two terms on the left side of Equation 7-3 are found in Row 1, Columns 3 and 4 and have values 10 and 0 respectively. According to Equation 7-3, these terms are summed to give $2S_2/\Delta t + O_2$, which is Column 5, Row 2 for the first time period. Therefore, the value 10 goes in that location.

But according to Figure 7-18, when $2S/\Delta t + O$ is 10, $O = 2$ cfs. Therefore, the value 2 goes in Column 6, Row 2. Figure 7-18 also shows that when $O = 2$ cfs, $2S/\Delta t - O = 6$ cfs. Therefore, the value 6 goes in Column 4, Row 2.

Now proceed to the second time period which starts at Row 2 and ends at Row 3. For this time period, the two terms on the left side of Equation 7-3 are found in Row 2, Columns 3 and 4 and have values 15 and 6 respectively. Therefore, the sum of these terms (15 + 6 = 21) goes in Column 5, Row 3. From Figure 7-18, when $2S/\Delta t + O = 21$ cfs, $O = 5$ cfs and $2S/\Delta t - O = 10$ cfs. These values are then written into the appropriate columns of Row 3.

Continue this procedure down the table and the values in Column 6 become the outflow hydrograph for the routing. Both inflow and outflow hydrographs are plotted in Figure 7-19. (Answer)

Table 7-2
Parameters Needed for Routing Computation in Example 7-5

O (cfs)	$\dfrac{2S}{\Delta t} - O$ (cfs)	$\dfrac{2S}{\Delta t} + O$ (cfs)
0.0	0.0	0.0
2.5	7.9	12.9
24.9	25.1	74.9
67.5	15.8	151
122	3.0	247

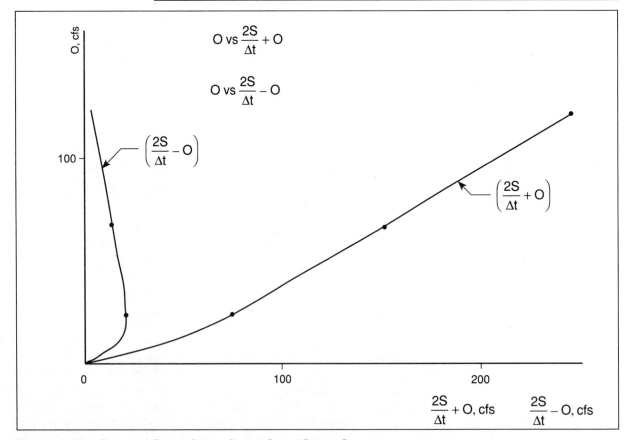

Figure 7-18 Graph of O vs $2S/\Delta t + O$ and O vs $2S/\Delta t - O$.

Table 7-3
Routing Table for Example 7-5

	(1) Time (hr)	(2) I_1 (cfs)	(3) $I_1 + I_2$ (cfs)	(4) $\frac{2S}{\Delta t} - O$ (cfs)	(5) $\frac{2S}{\Delta t} + O$ (cfs)	(6) O_2 (cfs)
(1)	0.0	4	10	0	0	—
(2)	.4	6	15	6	10	2
(3)	.8	9	32	10	21	5
(4)	1.2	23	107	18	42	11
(5)	1.6	84	132	20	125	53
(6)	2.0	48	68	15	152	68
(7)	2.4	20	34	25	83	30
(8)	2.8	14	25	22	59	18
(9)	3.2	11	20	19	47	12
(10)	3.6	9	17	16	39	9
(11)	4.0	8	15	15	33	8
(12)	4.4	7	14	12	30	7
(13)	4.8	7	13	10	26	5

A close look at Figure 7-19 reveals several facts about the routing in Example 7-5. First, notice that the maximum outflow was plotted as approximately 69 cfs even though the maximum O_2 value of the routing is 68 cfs. This is because the sketch reveals that the maximum outflow occurs between computed points.

Next, note that the maximum outflow occurs at the point where the outflow hydrograph crosses the inflow hydrograph. This is a characteristic of reservoir routings.

Figure 7-19   Inflow and outflow hydrographs for Example 7-5.

Also note that the maximum outflow of 69 cfs corresponds to a maximum impoundment elevation of 105.03 which is determined by interpolating the discharge rating in Figure 7- 17. This elevation would then become the design water level in the detention basin.

Finally, note that the routing resulted in a reduction of peak runoff from 84 cfs (inflow) to 69 cfs (outflow). This is an attenuation of 18 percent.

Example 7-5 illustrates one way to solve Equation 7-3 and perform a reservoir routing by hand. In practice, however, reservoir routings are rarely if ever computed by hand; instead, computers employing any of several available software programs are used. Refer to Appendix E for the names of applicable software.

The two detention basin case studies that follow utilize computer software to perform the routings and most of the other computations involved in detention basin design.

7.6 CASE STUDY 1

This case study focuses on the same subdivision located in Atlanta, Georgia, and discussed in Section 5.6 in relation to storm sewer design. The subdivision, called Tall Pines, is depicted in Figure 5-20. After completing storm sewer design, a detention facility was then designed to complete the stormwater management aspect of the project.

Design Requirements

According to local ordinance, the detention basin design criteria were:

1. Total peak runoff leaving the tract after development cannot exceed total peak runoff leaving the tract before development for the 2-year, 25-year and 100-year storms, each considered separately, and
2. Method of computation and design parameters are to be in accordance with good engineering practice.

The Modified Rational Method was chosen to be the method of computation since the project is relatively small and uncomplicated. Computations were performed on a personal computer using hydrologic software called Hydro Plus III. The program was used to compute the inflow hydrographs and the detention basin routings.

However, before starting the computing, an overall outline was devised. Since pre-development runoff traveled on the site in two general directions, northerly and southerly, the design process was made a little more involved than it might have been. Peak runoff under pre-development conditions would have to be computed in two parts, one for all runoff flowing in a northerly direction and one for all runoff flowing south. Figure 7-20 shows the tract under pre- development conditions divided into the two oppositely directed drainage areas.

Runoff Computations—Pre-Development

Peak runoff for each area was computed using the Rational Method in the usual way except that the drainage areas were not delineated in the usual way. The drainage areas were defined by tract boundaries and not by contour lines as described in Section 4.5. This results in artificial drainage areas but is accepted practice in determining peak runoff from a tract of land whose boundary does not coincide with a natural drainage basin.

Hydraulic path was delineated as starting at the remotest point in the drainage area and traveling to the farthest downhill point. This procedure also is slightly different from that in Section 4.5 because in this case we do not have a fixed point of analysis. Runoff crosses tract boundaries at several points, but only one is chosen as the point of analysis, namely the one that results in the longest hydraulic path.

Figure 7-20   Map of Tall Pines subdivision, existing conditions. *(Courtesy of Jaman Engineering.)*

Therefore, peak runoff under pre-development conditions was computed as follows:

A. Southerly Area
      A = 6.23 acres (measured from plan)
      c = 0.20 (entirely wooded tract)
      t_c = 17 min. (see hydraulic path)

Storm (years)	i (in/h)	Q_p (cfs)
2	3.2	4.0
25	5.7	7.1
100	7.0	8.7

Values of i were determined using the I-D-F curves for Atlanta in Appendix C-3.
Values of Q_p were determined using Equation 4-2.

B. Northerly Area
      A = 22.27 acres (measured from plan)
      c = 0.20 (entirely wooded tract)
      t_c = 25 min. (see hydraulic path)

Storm (years)	i (in/h)	Q_p (cfs)
2	2.7	12.0
25	4.8	21.4
100	5.8	25.8

Total peak runoff leaving the tract for pre-development conditions was then taken as the sum of the southerly and northerly runoffs.

Storm (years)	Q_p (cfs)
2	16.0
25	28.5
100	34.5

These values of Q_p then became the maximum allowable values of peak runoff under post-development conditions.

The location of the detention basin was chosen near the northerly end of the tract as shown in Figure 7-21 because this is a low point and because most of the improved areas drain to this point. Another reason was the proximity of the stream to receive outflow.

The detention basin was designed as an open cut basin because sufficient land area was available and open cut is much less costly than underground pipes.

Runoff Computations—Post-Development

Figure 7-21 shows that the developed tract is divided into three drainage areas,

1. Runoff flowing southerly,
2. Runoff tributary to the detention basin and
3. Runoff flowing northerly but bypassing the detention basin.

Figure 7-21   Map of Tall Pines subdivision, proposed conditions. *(Courtesy of Jaman Engineering.)*

Total post-development peak runoff is the sum of the peak runoffs from these three Areas. Runoff from areas 1 and 3 is computed just as was existing runoff while runoff from area 2 is equal to outflow from the detention basin.

Peak runoff from all three areas increases as a result of development because impervious areas cause the c-value to increase and reduced time of concentration causes the i-value to increase. Since Area 2 is the only area which can have reduced peak runoff (because of the detention basin) it is possible that Areas 1 and 3 could increase sufficiently to exceed total pre-development runoff on their own without the contribution of Area 2. If this happened, the detention basin would become inadequate because even if it reduced Area 2 runoff to zero, post-development runoff would still exceed pre-development runoff.

Unfortunately, this is exactly what happened when peak runoff was computed for Areas 1 and 3. This presented a major problem which had to be solved in order to continue with the project. Possible solutions were listed as:

1. Eliminate some of the proposed houses.
2. Re-grade the development to cause more lots to drain to the detention basin; that is, increase the size of Area 2 and reduce Areas 1 and 3.
3. Include seepage pits on each lot to intercept runoff from the proposed roofs. This would, in effect, eliminate the roofs from the drainage areas.

Solution 1 was held aside as a last resort. Solution 2 was considered too costly because of the great amount of earthwork involved. Solution 3 proved to be effective because it leads to a reduction of peak runoff from Areas 1 and 3 just sufficient to bring the sum of the two below the total pre-development peak runoff.

Therefore, two seepage pits were proposed for each lot and the area of the proposed roofs was subtracted from drainage area values in computing peak runoff.

Peak runoff under post-development conditions was computed as follows:

1. Southerly Area

$A = 6.88$ acres   (measured from map minus nine roofs)

$c = 0.33$
  Impervious —0.66 acres @ $c = 0.9$
  Lawns   —2.26 acres @ $c = 0.3$
  Woods   —3.95 acres @ $c = 0.2$

$t_c = 15.0$ min.   (see storm sewer design, Section 5.6)

Storm (years)	i (in/h)	Q_p (cfs)
2	3.5	7.2
25	6.0	12.4
100	7.2	14.9

2. Tributary to detention basin

$A = 9.46$ acres   (measured from map minus eleven roofs)

$c = 0.33$
  Impervious —1.31 acres @ $c = 0.9$
  Lawns   —2.93 acres @ $c = 0.3$
  Woods   —5.22 acres @ $c = 0.2$

$t_c = 25.7$ min. t_c to inlet 7 = 16.9 min.
  Pipe segment 7-8, t = 0.3 min.
  Swale 250' @ 1.0%,
    v = 1.1 fps, t = 3.8 min.
  Swale 420' @ 2.1%,
    v = 2.1 fps, t = 4.7 min.
      total = 25.7 min.

Storm (years)	i (in/h)	Q_p (cfs)
2	2.7	8.4
25	4.8	15.0
100	5.7	18.1

Note: The hydrograph of runoff from Area 2 was computed by Hydro Plus III software for basin routing purposes. See Figure 7-22.

3. Northerly area bypassing detention basin

    A = 11.25 acres (measured from map minus four roofs)

    c = 0.21   Impervious —0.11 acres @ c = 0.9

    Lawns   —0.54 acres @ c = 0.3

    Woods   —10.6 acres @ c = 0.2

    t_c = 15.6 min.   (see hydraulic path)

Storm (years)	i (in/h)	Q_p (cfs)
2	3.3	7.8
25	6.0	14.2
100	7.1	16.8

Total peak runoff under post-development conditions (before taking into account the detention basin) can be summarized for the three design storms as follows:

Storm (years)	Q_p (cfs) Area 1	Area 2	Area 3	Total Q_p (cfs)
2	7.2	8.4	7.8	23.4
25	12.4	15.0	14.2	41.6
100	14.9	18.1	16.8	49.8

These values are, as expected, greater than corresponding pre-development runoff values but they can be reduced to pre- development values by reducing Area 2 values through use of the detention basin.

For instance, if the 2-year runoff for Area 2 is reduced from 8.4 cfs to 1.0 cfs, then total post-development peak runoff would be reduced to 16.0 cfs which matches pre-development peak runoff. Similar reductions can be made for the 25-year and 100-year storms.

Detention Basin Computations

Therefore, the next step was to choose a detention basin size and compute the appropriate routings to determine peak outflow. After a few trial routings were performed, a detention basin as depicted in Figure 7-23 was selected. Parameters for this basin are listed in Figure 7-24.

Routing computations for this detention basin are shown in Figure 7-25. Notice that the column headings are similar to those shown in Table 7-3 used in Example 7-5. In Column 1, the parameter T is used for Time and is measured in minutes. Columns 2 and 3 represent inflow hydrograph values at the beginning and end of each time period and are measured in cfs. The next column lists values of 2S1/T measured in cfs which represents the parameter $2S_1/\Delta t$ from Equation 7-3. The next column, labeled O1, represents the parameter O_1 from Equation 7-3. The next column represents the parameter

HYDROLOGIC REPORT FOR

TALL PINES

UNIVERSAL RATIONAL HYDROGRAPH

Q(PEAK) = C*I*A
2 YEAR STORM FREQUENCY

BASIN IDENTIFIER DISCHARGES INTO	PROPOSED CONDITIONS DETENTION BASIN

BASIN AREA	=	9.46 ACRES
RUNOFF COEFF.	=	0.33
RAINFALL INT.	=	2.70 IN/HR
TIME OF CONC.	=	25.70 MINUTES
VOLUME	=	26811.27 CUBIC FEET

TIME (MIN)	RUNOFF (C.F.S.)
0.0	0.0
12.9	0.4
25.7	0.8
38.6	1.0
51.4	1.1
64.3	4.8
77.1	8.4
89.9	5.6
102.8	2.7
115.6	2.6
128.5	2.6
141.4	1.7
154.2	0.9
167.1	0.7
179.9	0.5
192.8	0.4
205.6	0.2
218.5	0.2
231.3	0.1
244.2	0.1
257.0	0.0
269.9	0.0
282.7	0.0
295.6	0.0
308.4	0.0
321.3	0.0
334.1	0.0
347.0	0.0
359.8	0.0
372.7	0.0

Figure 7-22. (Chart 1) Computer print-out of runoff hydrograph for post-development conditions tributary to detention basin. (2-year storm)

HYDROLOGIC REPORT FOR

TALL PINES

UNIVERSAL RATIONAL HYDROGRAPH

Q(PEAK) = C*I*A
25 YEAR STORM FREQUENCY

| BASIN IDENTIFIER | PROPOSED CONDITIONS |
| DISCHARGES INTO | DETENTION BASIN |

BASIN AREA	=	9.46 ACRES
RUNOFF COEFF.	=	0.33
RAINFALL INT.	=	4.80 IN/HR
TIME OF CONC.	=	25.70 MINUTES
VOLUME	=	56812.66 CUBIC FEET

TIME (MIN)	RUNOFF (C.F.S.)
0.0	0.0
12.9	1.6
25.7	3.1
38.6	3.1
51.4	3.1
64.3	9.1
77.1	15.0
89.9	9.1
102.8	3.1
115.6	3.1
128.5	3.1
141.4	3.1
154.2	3.1
167.1	3.1
179.9	3.1
192.8	3.1
205.6	3.1
218.5	1.6
231.3	0.0
244.2	0.0
257.0	0.0
269.9	0.0
282.7	0.0
295.6	0.0
308.4	0.0
321.3	0.0
334.1	0.0
347.0	0.0
359.8	0.0
372.7	0.0

Figure 7-22. (Chart 2) Computer print-out of runoff hydrograph for post-development conditions tributary to detention basin. (25-year storm)

HYDROLOGIC REPORT FOR

TALL PINES

UNIVERSAL RATIONAL HYDROGRAPH

Q(PEAK) = C*I*A
100 YEAR STORM FREQUENCY

BASIN IDENTIFIER DISCHARGES INTO	PROPOSED CONDITIONS DETENTION BASIN

BASIN AREA	=	9.46 ACRES
RUNOFF COEFF.	=	0.33
RAINFALL INT.	=	5.70 IN/HR
TIME OF CONC.	=	25.70 MINUTES
VOLUME	=	61145.09 CUBIC FEET

TIME (MIN)	RUNOFF (C.F.S.)
0.0	0.0
12.9	1.6
25.7	3.1
38.6	3.1
51.4	3.1
64.3	10.5
77.1	17.8
89.9	10.5
102.8	3.1
115.6	3.1
128.5	3.1
141.4	3.1
154.2	3.1
167.1	3.1
179.9	3.1
192.8	3.1
205.6	3.1
218.5	1.6
231.3	0.0
244.2	0.0
257.0	0.0
269.9	0.0
282.7	0.0
295.6	0.0
308.4	0.0
321.3	0.0
334.1	0.0
347.0	0.0
359.8	0.0
372.7	0.0

Figure 7-22. (Chart 3)   Computer print-out of runoff hydrograph for post-development conditions tributary to detention basin. (100-year storm)

Figure 7-23 Plan of proposed detention basin for Tall Pines subdivision. *(Courtesy of Jaman Engineering.)*

$2S_2/\Delta t + O_2$ from Equation 7-3. The next column, labeled O2, represents the parameter O_2 from Equation 7-3 and also becomes values of the outflow hydrograph. The last column lists values of 2S2/T representing the parameter $2S_2/\Delta t$ from Equation 7-3.

The time period, Δt, used in the routing is 12.85 min. which gives six points on the rising limb of the inflow hydrograph. It is the same time period used for the runoff hydrographs shown in Figure 7-22.

The routing computations show that the detention basin, as chosen, does a very good job of attenuating the inflow hydrograph. For the 100-year storm, peak inflow of 18.1 cfs is reduced to peak outflow of 1.9 cfs, a 90% attenuation. Similar attenuations are realized for the other two design storms.

HYDROLOGIC REPORT FOR

TALL PINES

POND DEPTH VS STORAGE PROVIDED

POND IDENTIFIER DETENTION BASIN

ELEV feet	DEPTH feet	AREA sq. ft.	VOLUME cu. ft.	SUM VOL cu. ft.
916.0		0.0		0.0
	1.0		3500.0	
917.0		7000.0		3500.0
	1.0		11750.0	
918.0		16500.0		15250.0
	1.0		18250.0	
919.0		20000.0		33500.0
	1.0		21000.0	
920.0		22000.0		54500.0
	1.0		23000.0	
921.0		24000.0		77500.0

ELEV	STORAGE (CU.FT.)	OUTFLOW (CFS)	2S/T+O (CFS)
916.0	0.0	0.0	0.0
917.0	3500.0	0.4	9.5
918.0	15250.0	0.6	40.1
919.0	33500.0	0.7	87.6
920.0	54500.0	2.5	143.9
921.0	77500.0	18.9	219.9

Figure 7-24 Computer printout showing detention basin parameters. Above is storage volume while below is a summary of storage volume and outflow rating.

The detention basin design summary can be stated by the following table:

Storm (years)	Total Peak Runoff (pre-development) (cfs)	Total Peak Runoff (post-development)			
		Area 1 (cfs)	Area 2* (cfs)	Area 3 (cfs)	Total (cfs)
2	16.0	7.2	0.6	7.8	15.6
25	28.5	12.4	1.7	14.2	28.3
100	34.5	14.9	1.9	16.8	33.6

* Peak outflow from detention basin

To satisfy the design criterium of zero increase in peak runoff, values in the last column above must be equal to or less than values in the second column. Since this is the case, the detention basin design was accepted.

HYDROLOGIC REPORT FOR

TALL PINES

HYDROGRAPH RESERVOIR ROUTING

    BASIN IDENTIFIER   PROPOSED CONDITIONS
    POND IDENTIFIER   DETENTION BASIN
    2 YEAR STORM FREQUENCY

T	I1	I2	2S1/T	O1	2S2/T +O2	O2	2S2/T
12.9	0.0	0.4	0.0	0.0	0.4	0.0	0.4
25.7	0.4	0.8	0.4	0.0	1.6	0.1	1.6
38.6	0.8	1.0	1.6	0.1	3.3	0.1	3.2
51.4	1.0	1.1	3.2	0.1	5.1	0.2	4.9
64.3	1.1	4.8	4.9	0.2	10.6	0.4	10.2
77.1	4.8	8.4	10.2	0.4	23.0	0.5	22.5
89.9	8.4	5.6	22.5	0.5	36.0	0.5	35.5
102.8	5.6	2.7	35.5	0.5	43.2	0.6	42.6
115.6	2.7	2.6	42.6	0.6	47.3	0.6	46.7
128.5	2.6	2.6	46.7	0.6	51.4	0.6	50.8
141.4	2.6	1.7	50.8	0.6	54.6	0.6	53.9
154.2	1.7	0.9	53.9	0.6	56.0	0.6	55.3
167.1	0.9	0.7	55.3	0.6	56.3	0.6	55.7
179.9	0.7	0.5	55.7	0.6	56.3	0.6	55.7
192.8	0.5	0.4	55.7	0.6	56.0	0.6	55.4
205.6	0.4	0.2	55.4	0.6	55.4	0.6	54.7
218.5	0.2	0.2	54.7	0.6	54.5	0.6	53.9
231.3	0.2	0.1	53.9	0.6	53.5	0.6	52.9
244.2	0.1	0.1	52.9	0.6	52.5	0.6	51.9
257.0	0.1	0.0	51.9	0.6	51.3	0.6	50.7
269.9	0.0	0.0	50.7	0.6	50.1	0.6	49.5
282.7	0.0	0.0	49.5	0.6	48.9	0.6	48.3
295.6	0.0	0.0	48.3	0.6	47.7	0.6	47.1
308.4	0.0	0.0	47.1	0.6	46.5	0.6	46.0
321.3	0.0	0.0	46.0	0.6	45.4	0.6	44.8
334.1	0.0	0.0	44.8	0.6	44.2	0.6	43.6
347.0	0.0	0.0	43.6	0.6	43.0	0.6	42.4
359.8	0.0	0.0	42.4	0.6	41.9	0.6	41.3
372.7	0.0	0.0	41.3	0.6	40.7	0.6	40.1

        MAXIMUM ELEVATION =   918.3 FT
        MAXIMUM STORAGE   = 21483.1 CU FT
        MAXIMUM DISCHARGE =   0.6 CFS

Figure 7-25. (Chart 1) Computer printout of detention basin routing computation (2-year storm).

HYDROLOGIC REPORT FOR

TALL PINES

HYDROGRAPH RESERVOIR ROUTING

BASIN IDENTIFIER   PROPOSED CONDITIONS
POND IDENTIFIER   DETENTION BASIN
25 YEAR STORM FREQUENCY

T	I1	I2	2S1/T	01	2S2/T +02	02	2S2/T
12.9	0.0	1.6	0.0	0.0	1.6	0.1	1.5
25.7	1.6	3.1	1.5	0.1	6.1	0.2	5.9
38.6	3.1	3.1	5.9	0.2	11.9	0.4	11.5
51.4	3.1	3.1	11.5	0.4	17.3	0.4	16.9
64.3	3.1	9.1	16.9	0.4	28.6	0.5	28.1
77.1	9.1	15.0	28.1	0.5	51.7	0.6	51.1
89.9	15.0	9.1	51.1	0.6	74.5	0.7	73.9
102.8	9.1	3.1	73.9	0.7	85.4	0.7	84.7
115.6	3.1	3.1	84.7	0.7	90.2	0.8	89.4
128.5	3.1	3.1	89.4	0.8	94.9	0.9	93.9
141.4	3.1	3.1	93.9	0.9	99.2	1.1	98.2
154.2	3.1	3.1	98.2	1.1	103.3	1.2	102.1
167.1	3.1	3.1	102.1	1.2	107.2	1.3	105.8
179.9	3.1	3.1	105.8	1.3	110.8	1.4	109.3
192.8	3.1	3.1	109.3	1.4	114.1	1.5	112.6
205.6	3.1	3.1	112.6	1.5	117.3	1.6	115.6
218.5	3.1	1.6	115.6	1.6	118.7	1.7	117.0
231.3	1.6	0.0	117.0	1.7	116.9	1.6	115.2
244.2	0.0	0.0	115.2	1.6	113.6	1.5	112.1
257.0	0.0	0.0	112.1	1.5	110.5	1.4	109.1
269.9	0.0	0.0	109.1	1.4	107.7	1.3	106.3
282.7	0.0	0.0	106.3	1.3	105.0	1.3	103.7
295.6	0.0	0.0	103.7	1.3	102.5	1.2	101.3
308.4	0.0	0.0	101.3	1.2	100.1	1.1	99.0
321.3	0.0	0.0	99.0	1.1	97.9	1.0	96.9
334.1	0.0	0.0	96.9	1.0	95.8	1.0	94.9
347.0	0.0	0.0	94.9	1.0	93.9	0.9	93.0
359.8	0.0	0.0	93.0	0.9	92.1	0.9	91.2
372.7	0.0	0.0	91.2	0.9	90.4	0.8	89.6

MAXIMUM ELEVATION =   919.6 FT
MAXIMUM STORAGE   = 45096.8 CU FT
MAXIMUM DISCHARGE =   1.7 CFS

Figure 7-25. (Chart 2)   Computer printout of detention basin routing computation (25-year storm).

HYDROLOGIC REPORT FOR

TALL PINES

HYDROGRAPH RESERVOIR ROUTING

    BASIN IDENTIFIER   PROPOSED CONDITIONS
    POND IDENTIFIER   DETENTION BASIN
    100 YEAR STORM FREQUENCY

T	I1	I2	2S1/T	O1	2S2/T +O2	O2	2S2/T
12.9	0.0	1.6	0.0	0.0	1.6	0.1	1.5
25.7	1.6	3.1	1.5	0.1	6.1	0.2	5.9
38.6	3.1	3.1	5.9	0.2	11.9	0.4	11.5
51.4	3.1	3.1	11.5	0.4	17.3	0.4	16.9
64.3	3.1	10.5	16.9	0.4	30.0	0.5	29.5
77.1	10.5	17.8	29.5	0.5	57.3	0.6	56.7
89.9	17.8	10.5	56.7	0.6	84.3	0.7	83.6
102.8	10.5	3.1	83.6	0.7	96.5	1.0	95.5
115.6	3.1	3.1	95.5	1.0	100.8	1.1	99.6
128.5	3.1	3.1	99.6	1.1	104.8	1.2	103.5
141.4	3.1	3.1	103.5	1.2	108.5	1.4	107.1
154.2	3.1	3.1	107.1	1.4	112.0	1.5	110.5
167.1	3.1	3.1	110.5	1.5	115.3	1.6	113.7
179.9	3.1	3.1	113.7	1.6	118.4	1.7	116.7
192.8	3.1	3.1	116.7	1.7	121.3	1.8	119.5
205.6	3.1	3.1	119.5	1.8	124.0	1.9	122.1
218.5	3.1	1.6	122.1	1.9	124.9	1.9	123.1
231.3	1.6	0.0	123.1	1.9	122.7	1.8	120.9
244.2	0.0	0.0	120.9	1.8	119.1	1.7	117.4
257.0	0.0	0.0	117.4	1.7	115.7	1.6	114.1
269.9	0.0	0.0	114.1	1.6	112.5	1.5	111.0
282.7	0.0	0.0	111.0	1.5	109.5	1.4	108.1
295.6	0.0	0.0	108.1	1.4	106.7	1.3	105.4
308.4	0.0	0.0	105.4	1.3	104.1	1.2	102.9
321.3	0.0	0.0	102.9	1.2	101.6	1.2	100.5
334.1	0.0	0.0	100.5	1.2	99.3	1.1	98.3
347.0	0.0	0.0	98.3	1.1	97.2	1.0	96.2
359.8	0.0	0.0	96.2	1.0	95.2	0.9	94.2
372.7	0.0	0.0	94.2	0.9	93.3	0.9	92.4

                MAXIMUM ELEVATION =   919.7 FT
                MAXIMUM STORAGE   = 47440.2 CU FT
                MAXIMUM DISCHARGE =   1.9 CFS

Figure 7-25. (Chart 3) Computer printout of detention basin routing computation (100-year storm).

At this point, two more design tasks remained: emergency spillway and erosion control.

Emergency Spillway

It was decided that due to the low potential downstream hazard of the detention basin that the emergency spillway could be incorporated into the outlet structure. Therefore, the top of the outlet structure, which functions as a weir, was designated as the emergency spillway.

The top of berm elevation was set one foot above the computed water level for the 100-year storm. This allows the peak runoff to enter the emergency spillway with a 1.0-foot freeboard remaining between the water and the top of berm. Figure 7-26 shows a schematic sketch of the outlet structure and emergency spillway.

Outfall

Notice that the outflow pipe was designed as a 24-inch concrete pipe. This is because the outflow pipe must be large enough that if full design flow entered the emergency spillway, the outflow pipe would have sufficient capacity to adequately convey the flow.

Finally, erosion protection at the outfall for the 24-inch outflow pipe was considered. To do this, first the exit velocity was determined for peak 100-year outflow of 1.9 cfs. (The emergency spillway outflow of 18.1 cfs was not used because of the remote possibility of this flow occurring.) Using Chart 39 in Appendix A-4, design velocity was found to be 3.7 fps. (The Manning's n value was assumed to be 0.012.)

Permissible velocity was checked in Appendix A-2 for earth channels with silt loam lining was found to be 3.0 fps. Since the design velocity exceeded the permissible, a riprap apron was proposed.

Figure 7-26 Schematic of detention basin outlet showing key design elevations.

To design the riprap lining, first tailwater depth was determined to be 0.38 feet using Chart 39 in Appendix A-4. Then, using Equation 5-1

$$d_{50} = \frac{.02}{.38} (1.9/2.0)^{4/3}$$

$$d_{50} = 0.049 \text{ ft (0.6 inches).}$$

Despite the result of 0.6-inch diameter, 2-inch diameter crushed stone, 6 inches thick was proposed to provide a more stable lining. Apron length was computed using Equation 5-2:

$$L_a = \frac{3(1.9)}{2.0^{3/2}}$$

$$L_a = 2.0 \text{ ft.}$$

Despite the minimal result of 2.0 ft, the apron length was proposed to be 5.0 ft to provide a more conservative design at minimal extra cost.

7.7 CASE STUDY 2

In this case study, an office site is proposed for a 28-acre tract in northwestern Missouri. The office development, called Liberty Road Associates, consists of a 4-story building with parking for 660 cars and is shown in Figure 7-27. A 900-foot long entrance drive sets the site well back from Martinsville Road on which the tract fronts.

A stream snakes close to the northwestern corner of the tract while another stream originates on site and flows northerly across the northern tract boundary. Most of the runoff leaving the site flows into the on-site stream.

Design Requirements

Stormwater management requirements for the development included an on-site detention basin designed in accordance with county regulations which included:

1. Total peak runoff leaving the tract after development cannot exceed total peak runoff leaving the tract before development for the 2, 5, 10, 25, 50 and 100-year storms, each considered separately.
2. Runoff and basin routing computations are to be in accordance with the SCS Method.
3. Emergency spillway design is to be based upon routing a storm with 10-inch, 24-hour rainfall.
4. Top of berm elevation is to be based upon routing a storm with 17-inch, 24-hour rainfall.

Computations for detention basin design were performed by the Hydro Plus III computer software on a personal computer.

Before starting the computations, a general design outline was devised. Terrain on the tract generally slopes downward from south to north with a high point located about 1000 feet south of the tract. The site is shown in relation to surrounding topography in Figure 7-28. The detention basin location was proposed along the northerly tract boundary as shown in Figure 7-29 because this is a low point on the site and is adjacent to the stream that flows off the site.

Figure 7-27   Topographic map showing the site layout for Liberty Road Associates office development. *(Courtesy of Canger & Cassera.)*

Watershed tributary
to detention basin

Tract
boundary

Topographic Map
Scale: 1" = 2000' (1:24000)
Contour interval 20'

Figure 7-28 Topographic map showing drainage area used for Liberty Road Associates detention design. *(Courtesy of U.S. Geological Survey.)*

The drainage area tributary to the proposed detention basin is delineated in Figure 7-28 and includes land uphill of the tract that drains onto the tract. This area was used to compute both pre-development and post-development runoff for detention design because it best represents total runoff leaving the tract. Runoff from the easterly portion of the tract upon which the access road was proposed was not included in the design because it was considered insignificant compared to the rest of the site. However, an analysis of the existing storm sewer system in Martinsville Road was performed to show that sufficient capacity existed to accommodate increased flow from the project access drive.

As shown in Figure 7-29, the detention basin was incorporated into the site as a landscaping feature integral to the building. A permanent pond was designed at the bottom of the basin to serve as a water enhancement for the site. Thus, the basin was designed as a wet detention basin. An internal site driveway was proposed on top of the detention basin berm so that the berm could serve two functions.

Figure 7-29 Plan of detention basin showing proposed contours. *(Courtesy of Canger & Cassera.)*

Runoff Computation—Pre-Development

To compute peak runoff under pre-development conditions, the drainage area used was the area tributary to the detention basin. Although this is a post-development area, it was used so that the pre- and post-development drainage areas would be equal and therefore give a fair comparison of runoff.

Figure 7-30 shows pre- and post-development watersheds tributary to the stream at the northerly tract boundary. Since the areas are not equal, they cannot be used for a fair comparison for runoff quantities. Therefore, the pre- development watershed delineation was changed to match the post-development delineation even though this resulted in an artificially drawn watershed.

Figure 7-30 Plan showing a comparison of drainage areas for existing and proposed conditions.

Hydraulic path was taken as starting at the remotest point of the watershed and travelling to the farthest downhill point similar to the method used for Case Study 1. As shown in Figure 7-29, this hydraulic path does not end at the stream but it does provide the longest time for runoff computation.

Therefore, peak runoff under pre-development conditions was computed by the SCS Method as follows:

A_m = 33.45 acres = 0.0523 s.m. (measured from plan)

Curve Number: From the local soil survey, soil types were identified and found to have Hydrologic Soil Group B. The entire watershed was wooded, so from Appendix D-1,

CN = 55.

Time of Concentration: From the delineated hydraulic path

t_c = 40 min.

Rainfall: Using the various maps in Appendix D-3, rainfall was found for each design storm for the project location.

Storm (years)	Rainfall (in.)
100	7.5
50	6.5
25	6.0
10	5.2
5	4.3
2	3.3

Rainfall distribution: From Appendix D-4, rainfall distribution was found to be Type II.

Peak runoff: Using the parameters above, the computer program computed peak runoff as follows

Storm (years)	q_p (cfs)
100	51.7
50	36.7
25	29.7
10	19.5
5	10.2
2	2.9

These values of peak runoff then became the maximum allowable values of peak runoff under post-development conditions.

Runoff Computations—Post-Development

Next, post-development runoff hydrographs were computed by the computer and stored for use in routing computations. SCS runoff parameters were

A_m = 33.45 acres = 0.0523 s.m. (measured from plan)

Curve Number: Soil cover types are impervious, landscaped and wooded
Impervious   8.5 acres   @ CN = 98
Landscaped 12.95 acres @ CN = 65
Wooded   12 acres   @ CN = 55

CN = 70.

Time of concentration: From the delineated hydraulic path in Figure 7-30, t_c consists of three components:
Overland flow   32 min.
Shallow concentrated flow   3 min.
Pipe flow   1.1 min.

t_c = 36 min.

Rainfall: Same as for pre-development runoff.

Rainfall distribution: Same as for pre-development runoff.

Peak runoff: Using the parameters above, the computer program computed runoff hydrographs having the following peak values:

Storm (years)	q_p (cfs)
100	98.9
50	78.1
25	68.0
10	52.4
5	35.7
2	19.1

These values are about double pre-development peak runoff (except the smaller storms which are as much as six times pre-development peak runoff). Why are the smaller storm runoffs so much more than corresponding pre-development runoffs?

Detention Basin Computations

By routing through the detention basin, the runoff rates were attenuated to rates below pre-development runoff values. The detention basin outflow structure is shown in Figure 7-31 and basin parameters are listed in Figure 7-32.

Figure 7-31   Outflow structure for Liberty Road Associates detention design.

The routings resulted in the following peak outflow rates and water levels:

(1) Storm (years)	(2) Allowable Outflow (cfs)	(3) Peak Inflow (cfs)	(4) Peak Outflow (cfs)	(5) Max. W.L. Elevation (ft.)
100	51.7	98.9	50.0	298.90
50	36.7	78.1	36.4	298.10
25	29.7	68.0	28.7	297.80
10	19.5	52.4	16.8	297.20
5	10.2	35.7	7.0	296.40
2	2.9	19.1	2.3	295.30

To satisfy the design criterium of zero increase in peak runoff, values in Column 4 above must be equal to or less than values in Column 2. Since this is the case, the detention basin design was accepted.

Inflow and outflow hydrographs for the 100-year storm are plotted in Figure 7-33. Hydrographs for the remaining storms have similar shapes.

A close inspection of Figure 7-31 reveals that the outflow structure is a 3-stage outflow structure. The first stage is a 6-inch orifice, the second stage is a 15-inch orifice and the third stage consists of two 24-inch orifices on both sides of the structure. The top

HYDROLOGIC REPORT FOR LIBERTY ROAD ASSOCIATES STAGE, STORAGE + DISCHARGE

POND IDENTIFIER PROP. DETENTION POND

1 =

ELEV	STORAGE (CU. FT.)	OUTFLOW (CFS)	2S/T+O (CFS)
293.0	0.0	0.0	0.0
293.5	9343.8	0.5	62.8
294.0	19875.0	0.9	133.4
294.5	31593.8	1.1	211.7
295.0	44500.0	1.3	298.0
295.5	58330.0	2.9	391.8
296.0	72816.3	5.4	490.8
296.5	87972.5	7.5	594.0
297.0	103785.0	13.3	705.2
297.5	120240.0	22.5	824.1
298.0	137351.3	33.5	949.2
298.5	155120.0	43.8	1077.9
299.0	173545.0	51.7	1208.7
300.0	212295.0	320.7	1736.0
301.0	253670.0	725.0	2416.1

Figure 7-32 Computer printout showing detention basin parameters for Liberty Road Associates site.

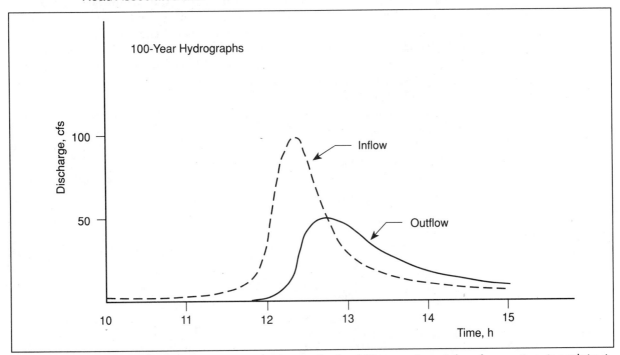

Figure 7-33 Plotting of inflow and outflow hydrographs for 100-year storm taken from computer printout.

of the structure contains an access manhole cover usually in the closed position. Therefore, the top does not serve hydraulically as another stage. Multiple stages were used because of the multiple storm routings. The outflow rating must be fashioned in such a way as to result in zero increase in runoff for each of the six storms. This can be done efficiently only by manipulating the outflow with multiple stages.

Emergency Spillway

The emergency spillway was incorporated into the berm by providing a low-point in the roadway. The elevation of the low point is 299.0, which is just above the maximum 100-year impoundment level of 298.90. Thus, when the water level in the detention basin rises 0.10 feet above the 100-year maximum level, it starts to flow over the emergency spillway.

Normally, an emergency spillway is not allowed to be incorporated as an overtopping of the berm unless sufficient reinforcement of the berm surface is provided. So in this case, the downstream side of the berm was stabilized so that emergency overtopping could take place without eroding the soil and causing a breach or failure of the berm.

A 17-inch storm was routed through the basin to determine freeboard, or the vertical distance from the emergency spillway crest to the top of berm. The routing resulted in a maximum water level elevation of 299.90. This elevation was used to determine the extent of stabilization of the downstream side of the berm.

This berm design was unconventional because the top elevation was varied along its extent, with higher elevations at each end and a low point near the outflow structure. Normally, the berm would be level and the emergency spillway would be a wide channel cut out of one end. However, in this case, such an arrangement would not have allowed a roadway to traverse the berm, and since the roadway was needed for internal site circulation, the unconventional overtopping proposal was approved.

It is important to note that the only way that the overtopping emergency spillway was made acceptable was by stabilizing the top of the berm with the paved roadway and stabilizing the downstream side with other materials.

Outfall

The last aspect of this detention basin was the selection of the outflow pipe and the outflow apron. The outflow pipe was designed to convey the 100-year outflow with a headwater elevation below the second stage, elevation 295.0. This was accomplished with a 42-inch concrete pipe with invert set at elevation 290.0 as shown in Figure 7-31. Treating the pipe as a culvert with inlet control, a 100-year discharge of 50.0 cfs resulted in HW = 3.43 feet, equivalent to a water elevation inside the outflow structure of 293.43, which is below the second stage.

Erosion protection at the outfall of the 42-inch outflow pipe was considered by computing exit velocity for the 100- year outflow of 50.0 cfs. Using Chart 44 in Appendix A-4, design velocity was found to be about 9.0 fps, with a tailwater depth, TW, of 1.65 feet. (The Manning's n value was assumed to be 0.012.)

The exit velocity of 9.0 fps is well above any permissible velocity found in Appendix A-2. Therefore, a riprap apron was designed. Using Equation 5-1

$$d_{50} = \frac{.02}{1.65} \left(\frac{50}{3.5} \right)^{4/3}$$

$$d_{50} = 0.42 \text{ ft} \quad (5 \text{ inches})$$

use $d_{50} = 6$ inches.

Apron length was computed using Equation 5-2

$$L_a = \frac{3(50)}{3.5^{3/2}}$$

$$L_a = 22.9 \text{ ft.}$$

Despite the computed length of 22.9 feet, the apron was proposed to be 20.0 feet long since that was the distance from the outfall headwall to the northerly property line.

PROBLEMS

1. Referring to the detention basin shown in Figure 7-34, compute a chart of storage volume values for elevations 100, 101, 102, 102.5, 104 and 106.

2. Compute the outflow rating for the detention basin shown in Figure 7-34. Use the same elevations listed in problem 1. Ignore the constriction caused by the outflow pipe.

3. Using the results of Problems 1 and 2, plot graphs of O vs 2S/Δt + O and O vs 2S/Δt − O.

4. Using the results of Problems 1 through 3, route the inflow hydrograph listed in Table 7-4 through the detention basin shown in Figure 7-34. From the routing determine peak outflow and maximum water level.

<table>
<tr><td colspan="2" align="center">**Table 7-4**
Inflow Hydrograph for Detention Basin Routing</td></tr>
<tr><td align="center">**Time**
(hr)</td><td align="center">**Inflow**
(cfs)</td></tr>
<tr><td align="center">0.0</td><td align="center">1.2</td></tr>
<tr><td align="center">.25</td><td align="center">1.5</td></tr>
<tr><td align="center">.50</td><td align="center">2.1</td></tr>
<tr><td align="center">.75</td><td align="center">3.2</td></tr>
<tr><td align="center">1.00</td><td align="center">11.6</td></tr>
<tr><td align="center">1.25</td><td align="center">22.1</td></tr>
<tr><td align="center">1.50</td><td align="center">9.8</td></tr>
<tr><td align="center">1.75</td><td align="center">4.5</td></tr>
<tr><td align="center">2.00</td><td align="center">3.2</td></tr>
<tr><td align="center">2.25</td><td align="center">2.7</td></tr>
<tr><td align="center">2.50</td><td align="center">2.4</td></tr>
<tr><td align="center">2.75</td><td align="center">2.0</td></tr>
<tr><td align="center">3.00</td><td align="center">1.8</td></tr>
</table>

5. Design an open-cut detention basin for the developed tract shown in Figure 7-35 to result in zero increase in peak runoff for the 100-year storm. Computations are to be by SCS Method, either by computer or by hand. Locate the basin on site south of the parking area and discharge at one of the on-site streams. Make all slopes three horizontal to one vertical.

Assumptions:

1. Project location: near St. Paul, Minnesota

2. Hydrologic soil group B

Figure 7-34 Plan and section of detention basin.

3. Pre-development soil cover on site: wooded, good condition
4. Post-development soil cover on site:
   impervious
   lawns, good condition
   woods, fair condition

Figure 7-35 Topographic map of site located near St. Paul, Minnesota. *(Map adapted from Aero Service.)*

FURTHER READING

Akan, A. Osman. (1993). *Urban Stormwater Hydrology, A Guide to Engineering Calculations*. Lancaster, PA: Technomic Publishing Co.

Debo, T., Reese, A. (1995). *Municipal Storm Water Management*. Boca Raton: Lewis Publishers.

Gupta, R. S. (1995). *Hydrology and Hydraulic Systems*. Prospect Heights, IL: Waveland Press, Inc.

U.S. Department of Agriculture. (1985). *National Engineering Handbook, Section 4, Hydrology*. Washington, D.C.: Soil Conservation Service.

The Urban Water Resources Research Council of ASCE and the Water Environment Federation. (1992). *Design and Construction of Urban Stormwater Management Systems*. New York, NY and Alexandria, VA: ASCE.

APPENDIX CONTENTS

A

DESIGN CHARTS FOR OPEN CHANNEL FLOW

A-1
Roughness Coefficients (Manning's n)

I. CLOSED CULVERTS:
    A. Concrete pipe .... 0.012 – 0.015
    B. Corrugated-metal pipe or pipe-arch (annular unpaved):
        1. $2^{2}/_{3}$ by $1/_{2}$-in. corrugation riveted pipe 0.024
        2. 3 in. by 1 in. corrugation 0.027
        3. 5 in. by 1 in. corrugation 0.025
        4. 6 in. by 2 in. corrugation (field bolted) 0.033
    C. Vitrified clay pipe .... 0.012 – 0.014
    D. Cast-iron pipe, uncoated 0.013
    E. Steel pipe .... 0.009 – 0.011
    F. Brick .... 0.014 – 0.017
    G. Monolithic concrete:
        1. Wood forms, rough .... 0.015 – 0.017
        2. Wood forms, smooth .... 0.012 – 0.014
        3. Steel forms .... 0.012 – 0.013
    H. Cemented rubble masonry walls:
        1. Concrete floor and top .... 0.017 – 0.022
        2. Natural floor .... 0.019 – 0.025
    I. Laminated treated wood .... 0.015 – 0.017
    J. Vitrified clay liner plates 0.015
    K. Polyvinyl Chloride PVC .... 0.007 – 0.011

II. OPEN CHANNELS, LINED (Straight Alignment):
    A. Concrete, with surfaces as indicated:
        1. Formed, no finish .... 0.013 – 0.017
        2. Trowel finish .... 0.012 – 0.014
        3. Float finish .... 0.013 – 0.015
        4. Float finish, some gravel on bottom .... 0.015 – 0.017
        5. Gunite, good section .... 0.016 – 0.019
        6. Gunite, wavy section .... 0.018 – 0.022
    B. Concrete, bottom float finished, sides as indicated:
        1. Dressed stone in mortar .... 0.015 – 0.017
        2. Random stone in mortar .... 0.017 – 0.020
        3. Cement rubble masonry .... 0.020 – 0.025
        4. Cement rubble masonry, plastered .... 0.016 – 0.020
        5. Dry rubble (riprap) .... 0.020 – 0.030
    C. Gravel bottom, sides as indicated:
        1. Formed concrete .... 0.017 – 0.020
        2. Random stone in mortar .... 0.020 – 0.023
        3. Dry rubble (riprap) .... 0.023 – 0.033
    D. Brick .... 0.014 – 0.017
    E. Asphalt:
        1. Smooth 0.013
        2. Rough 0.016
    F. Wood, planed, clean .... 0.001 – 0.013
    G. Concrete-lined excavated rock:
        1. Good section .... 0.017 – 0.020
        2. Irregular section .... 0.022 – 0.027

III. OPEN CHANNELS, EXCAVATED (Straight Alignment, Natural Lining):
    A. Earth, uniform section:
        1. Clean, recently completed .... 0.016 – 0.018
        2. Clean, after weathering .... 0.018 – 0.020
        3. With short grass, few weeds .... 0.022 – 0.027
        4. In gravelly soil, uniform section, clean .... 0.022 – 0.025
    B. Earth, fairly uniform section:
        1. No vegetation .... 0.022 – 0.025
        2. Grass, some weeds .... 0.025 – 0.030
        3. Dense weeds or aquatic plants in deep channels .... 0.030 – 0.035
        4. Sides clean, gravel bottom .... 0.025 – 0.030
        5. Sides clean, cobble bottom .... 0.030 – 0.040

C. Dragline excavated or dredged:
    1. No vegetation .... 0.028 – 0.033
    2. Light brush on banks .... 0.035 – 0.050
D. Rock
    1. Based on design section 0.035
    2. Based on actual mean section:
       a. Smooth and uniform .... 0.035 – 0.040
       b. Jagged and irregular .... 0.040 – 0.045
E. Channels not maintained, weeds and brush uncut:
    1. Dense weeds, high as flow depth.............................. 0.080 – 0.120
    2. Clean bottom, brush on sides .... 0.050 – 0.080
    3. Clean bottom, brush on sides, highest stage of flow .... 0.070 – 0.140
    4. Dense brush, high stage .... 0.100 – 0.140

IV. HIGHWAY CHANNELS AND SWALES WITH MAINTAINED VEGETATION
(Values shown are for velocities of 2 and 6 f.p.s.):
A. Depth of flow up to 0.7 foot:
    1. Bermudagrass, Kentucky bluegrass, buffalograss:
       a. Mowed to 2 inches .... 0.070 – 0.045
       b. Length 4 to 6 inches .... 0.090 – 0.050
    2. Good stand, any grass:
       a. Length about 12 inches .... 0.180 – 0.090
       b. Length about 24 inches .... 0.200 – 0.100
    3. Fair stand, any grass:
       a. Length about 12 inches .... 0.140 – 0.080
       b. Length about 24 inches .... 0.250 – 0.130
B. Depth of flow 0.7–1.5 feet:
    1. Bermudagrass, Kentucky bluegrass, buffalograss:
       a. Mowed to 2 inches .... 0.050 – 0.035
       b. Length 4 to 6 inches .... 0.060 – 0.040
    2. Good stand, any grass:
       a. Length about 12 inches .... 0.120 – 0.070
       b. Length about 24 inches .... 0.020 – 0.100
    3. Fair stand, any grass:
       a. Length about 12 inches .... 0.100 – 0.060
       b. Length about 24 inches .... 0.170 – 0.090

V. STREET AND EXPRESSWAY GUTTERS:
A. Concrete gutter, troweled finish 0.012
B. Asphalt pavement:
    1. Smooth texture 0.013
    2. Rough texture 0.016
C. Concrete gutter with asphalt pavement:
    1. Smooth 0.013
    2. Rough 0.015
D. Concrete pavement:
    1. Float finish 0.014
    2. Broom finish 0.016
E. For gutters with small slope, where sediment may accumulate, increase above values of n by 0.002

VI. NATURAL STREAM CHANNELS:
A. Minor streams (surface width at flood stage less 100 ft.):
    1. Fairly regular section:
       a. Some grass and weeds, little or no brush .... 0.030 – 0.035
       b. Dense growth of weeds, depth of flow materially greater than weed height .... 0.030 – 0.050
       c. Some weeds, light brush on banks .... 0.035 – 0.050
       d. Some weeds, heavy brush on banks .... 0.050 – 0.070
       e. Some weeds, dense willows on banks .... 0.060 – 0.080
       f. For trees within channel, with branches submerged at high stage, increase all
          above values by .... 0.010 † 0.020
    2. Irregular sections, with pools, slight channel meander; increase values given
       in 1.a.–e. about .... 0.010 – 0.020
    3. Mountain streams, no vegetation in channel banks usually steep, trees and brush
       along banks submerged at high stage:
       a. Bottom of gravel, cobbles, and few boulders .... 0.040 – 0.050
       b. Bottom of cobbles, with large boulders .... 0.050 - 0.070

B. Flood plains (adjacent to natural streams):
  1. Pasture, no brush:
    a. Short grass .... 0.030 − 0.035
    b. High grass .... 0.035 − 0.050
  2. Cultivated areas:
    a. No crop .... 0.030 − 0.040
    b. Mature row crops .... 0.035 − 0.045
    c. Mature field crops .... 0.040 − 0.050
  3. Heavy weeds, scattered brush .... 0.050 − 0.070
  4. Light brush and trees:
    a. Winter .... 0.050 − 0.060
    b. Summer .... 0.060 − 0.080
  5. Medium to dense brush:
    a. Winter .... 0.070 − 0.110
    b. Summer .... 0.100 − 0.170

(Courtesy of New Jersey Dept. of Transportation, Design, Manual, Roadway)

A-2
Permissible Velocities

SOIL TEXTURE	ALLOWABLE VELOCITY (ft/sec)
Sand and sandy loam (noncollodial)	2.5
Silt loam (also high lime clay)	3.0
Sandy clay loam	3.5
Clay loam	4.0
Clay, fine gravel, graded loam to gravel	5.0
Cobbles	5.5
Shale	6.0

(Courtesy of New Jersey State Soil Conservation Committee, Standards for Soil Erosion and Sediment Control in New Jersey.)

A-3
Channel Charts for the Solution of Manning's Equation

The following selected design charts provide a direct graphical solution of Manning's Equation for various sized open channels with rectangular and trapezoidal cross-section. The channels are assumed to have uniform slope, cross-section, and roughness and are not affected by backwater.

Charts 1 through 14 (rectangular channels) are calibrated for a roughness value of 0.015, and Charts 15 through 23 (trapezoidal channels) are calibrated for a roughness value of 0.030. However, by use of the Qn scale, any n value may be used. Use of the Qn scale is explained in Item 1 below.

You may use the charts to find discharge, velocity, normal depth, and critical depth. Following is a brief outline describing basic use of the charts:

1. **How to use the Qn scale**—If n = 0.015 for a rectangular channel or n = 0.030 for a trapezoidal channel, do not use the Qn scale, but read the discharge directly on the Q scale. However, for all other n values, multiply n by the discharge, Q, and read the resulting Qn value on the Qn scale directly below the Q scale.

2. **How to find normal depth**—If you know discharge, roughness and slope, find normal depth by entering the value of Q (or Qn) and project a line straight up the graph until you intersect the appropriate slope line. Using the point of intersection, read the normal depth using the series of diagonal lines labeled *NORMAL DEPTH OF FLOW—FEET*. You may need to interpolate between lines.

3. **How to find critical depth**—Critical depth is independent of roughness, so do not use the Qn scale. If you know the discharge and slope, enter the value of Q and project straight up until you intersect the *CRITICAL* curve (dashed line). Using this point of intersection, read normal depth as in Item 2 above. The resulting value of normal depth is critical depth. (Remember that critical depth is a normal depth as well.)

4. **How to find velocity**—If you know discharge, roughness, and slope, find velocity by entering the value of Q (or Qn) and project a line straight up until you intersect the appropriate slope line. Using the point of intersection, read the velocity using the series of curves running down and to the left labeled *VELOCITY—FPS*. You may need to interpolate between lines.

5. **How to find discharge**—If you know slope, normal depth, and roughness, find discharge by first locating the intersection of the appropriate slope and normal depth lines (interpolating if necessary). From the point of intersection, project a line straight down to the Qn scale and read the value of Qn. Finally, divide the Qn value by n to obtain discharge.

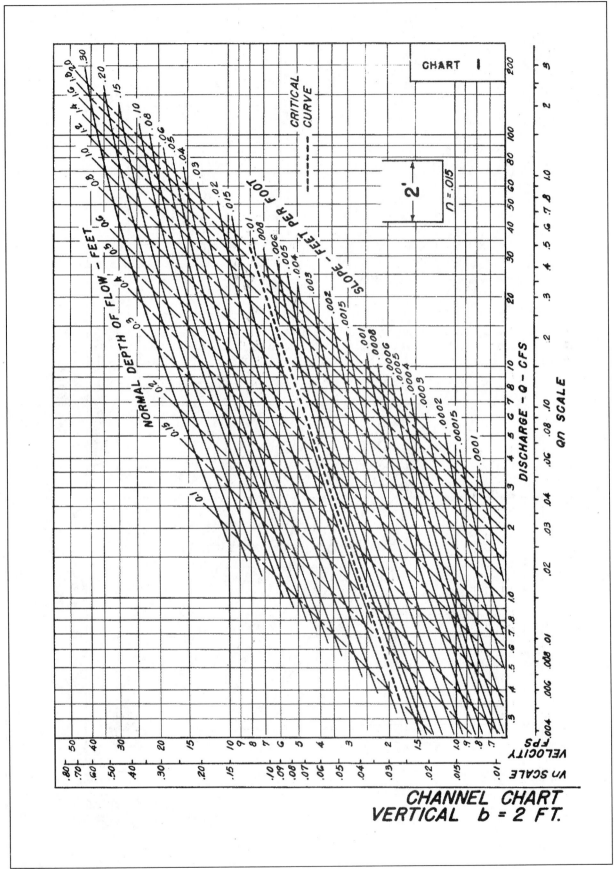

Figure A-3. (Chart 1)   Channel charts for the solution of Manning's Equation. *(Courtesy of U.S. Dept. of Commerce, Bureau of Public Roads, Design Charts for Open-Channel Flow.)*

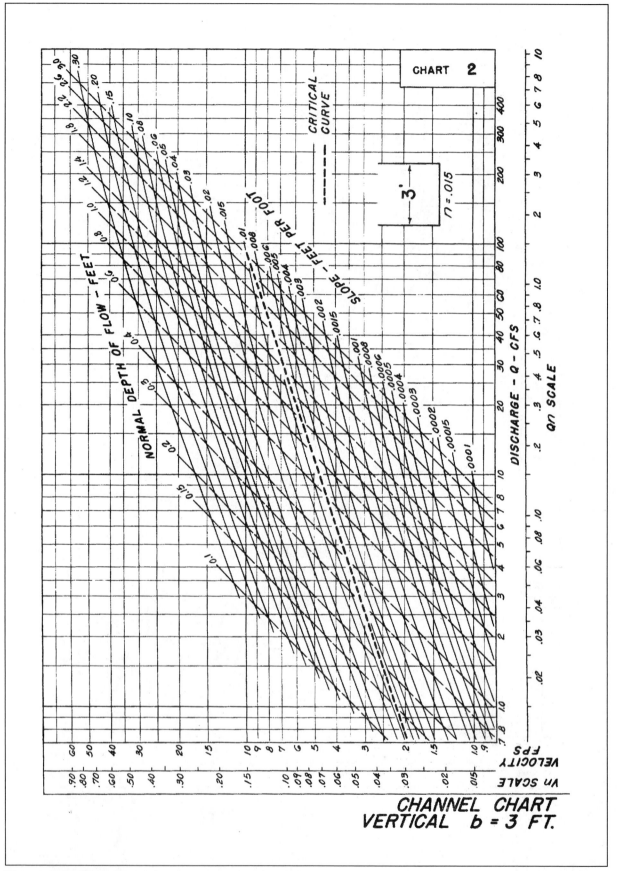

Figure A-3. (Chart 2) Channel charts for the solution of Manning's Equation. *(Courtesy of U.S. Dept. of Commerce, Bureau of Public Roads, Design Charts for Open-Channel Flow.)*

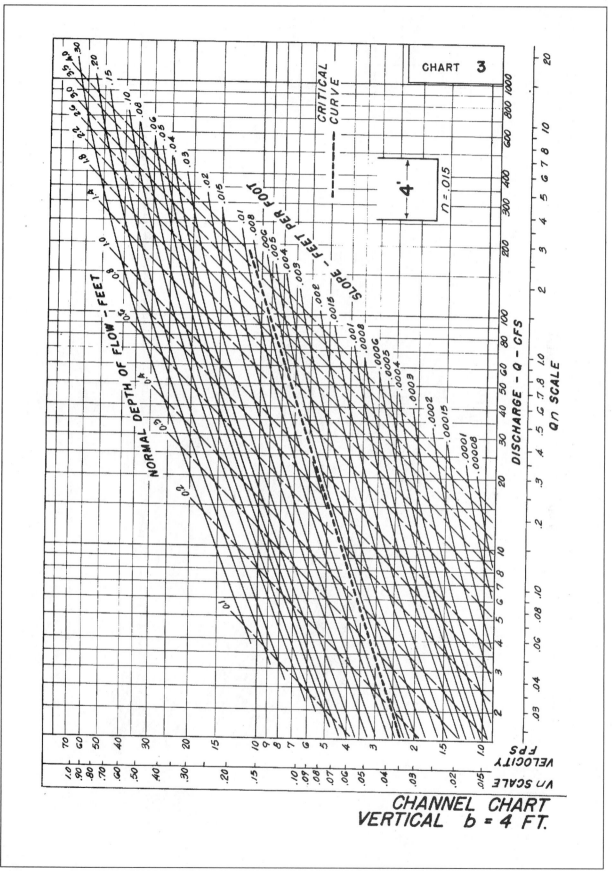

Figure A-3. (Chart 3) Channel charts for the solution of Manning's Equation. *(Courtesy of U.S. Dept. of Commerce, Bureau of Public Roads, Design Charts for Open-Channel Flow.)*

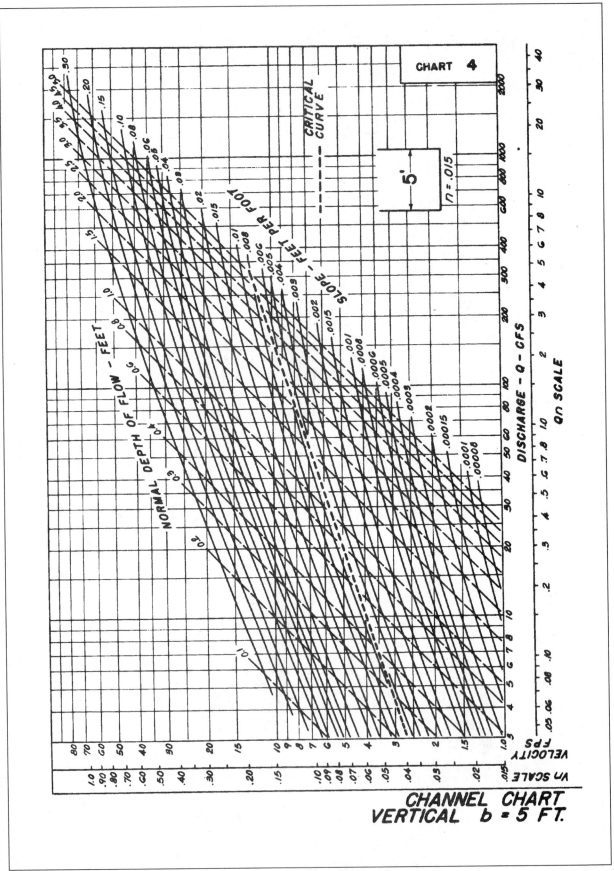

Figure A-3. (Chart 4) Channel charts for the solution of Manning's Equation. *(Courtesy of U.S. Dept. of Commerce, Bureau of Public Roads, Design Charts for Open-Channel Flow.)*

Figure A-3. (Chart 5)   Channel charts for the solution of Manning's Equation. *(Courtesy of U.S. Dept. of Commerce, Bureau of Public Roads, Design Charts for Open-Channel Flow.)*

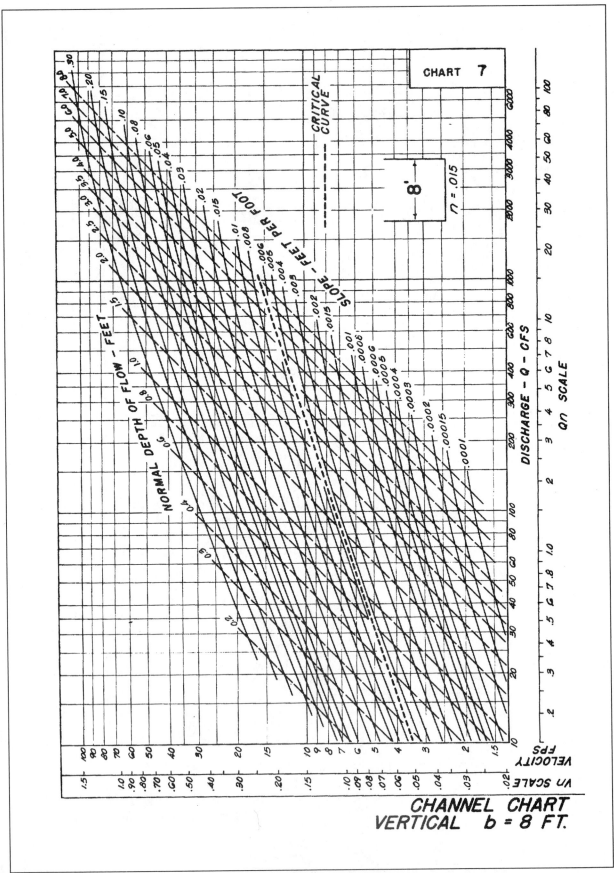

Figure A-3. (Chart 7) Channel charts for the solution of Manning's Equation. *(Courtesy of U.S. Dept. of Commerce, Bureau of Public Roads, Design Charts for Open-Channel Flow.)*

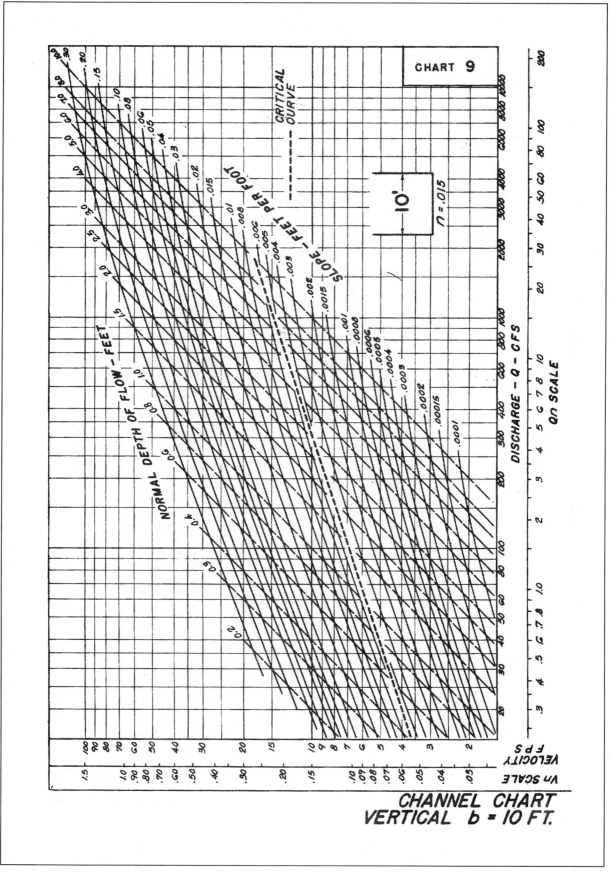

Figure A-3. (Chart 9)   Channel charts for the solution of Manning's Equation. *(Courtesy of U.S. Dept. of Commerce, Bureau of Public Roads, Design Charts for Open-Channel Flow.)*

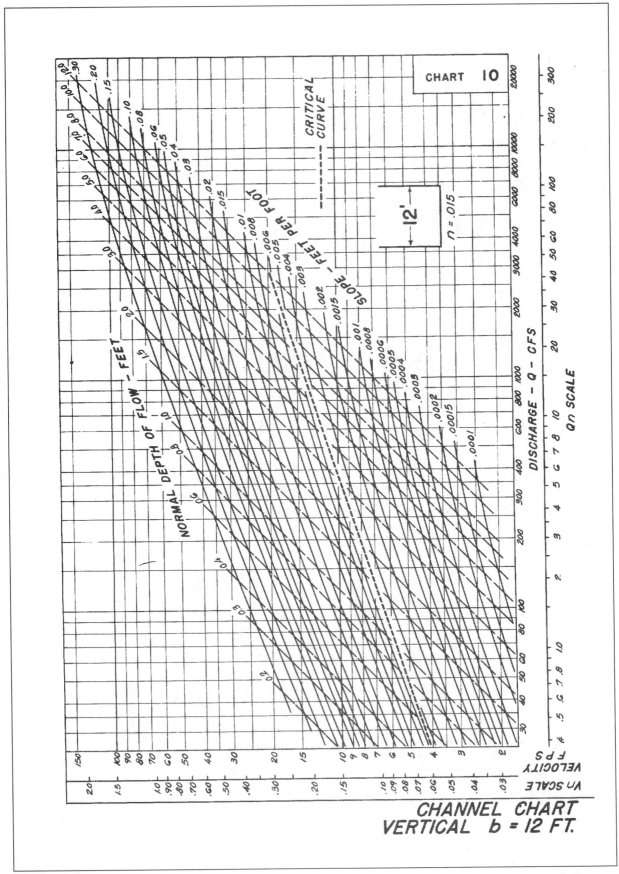

Figure A-3. (Chart 10) Channel charts for the solution of Manning's Equation. *(Courtesy of U.S. Dept. of Commerce, Bureau of Public Roads, Design Charts for Open-Channel Flow.)*

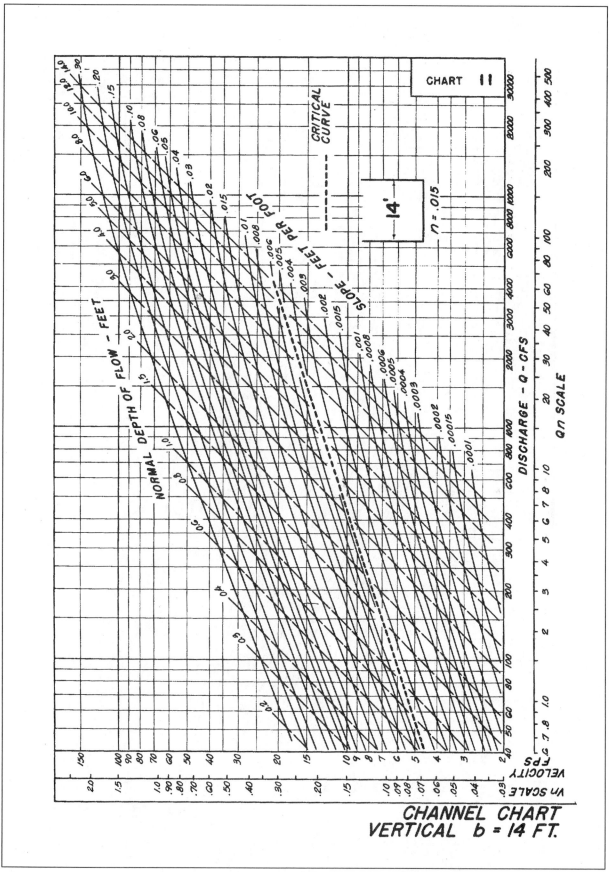

Figure A-3. (Chart 11)   Channel charts for the solution of Manning's Equation. *(Courtesy of U.S. Dept. of Commerce, Bureau of Public Roads, Design Charts for Open-Channel Flow.)*

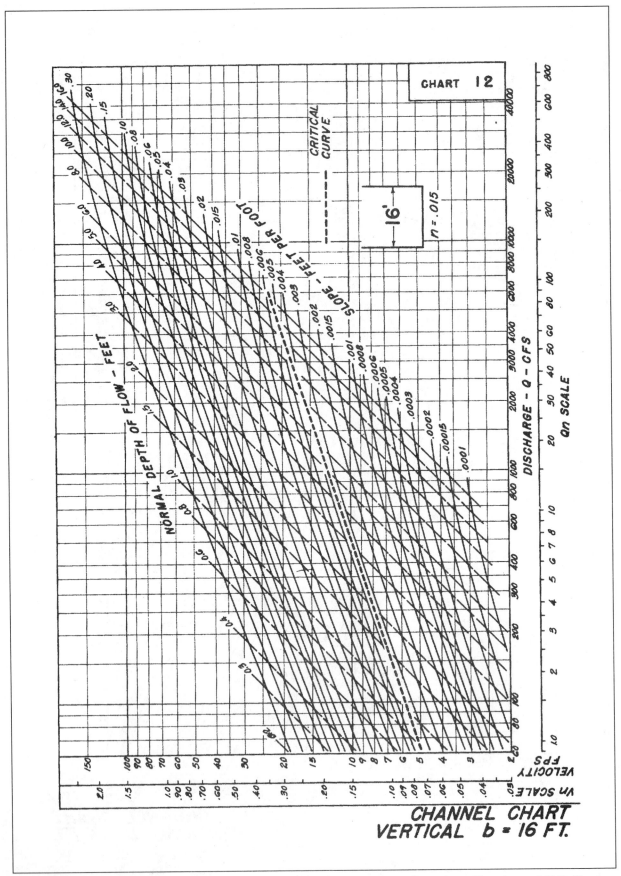

Figure A-3. (Chart 12) Channel charts for the solution of Manning's Equation. *(Courtesy of U.S. Dept. of Commerce, Bureau of Public Roads, Design Charts for Open-Channel Flow.)*

Figure A-3. (Chart 13) Channel charts for the solution of Manning's Equation. *(Courtesy of U.S. Dept. of Commerce, Bureau of Public Roads, Design Charts for Open-Channel Flow.)*

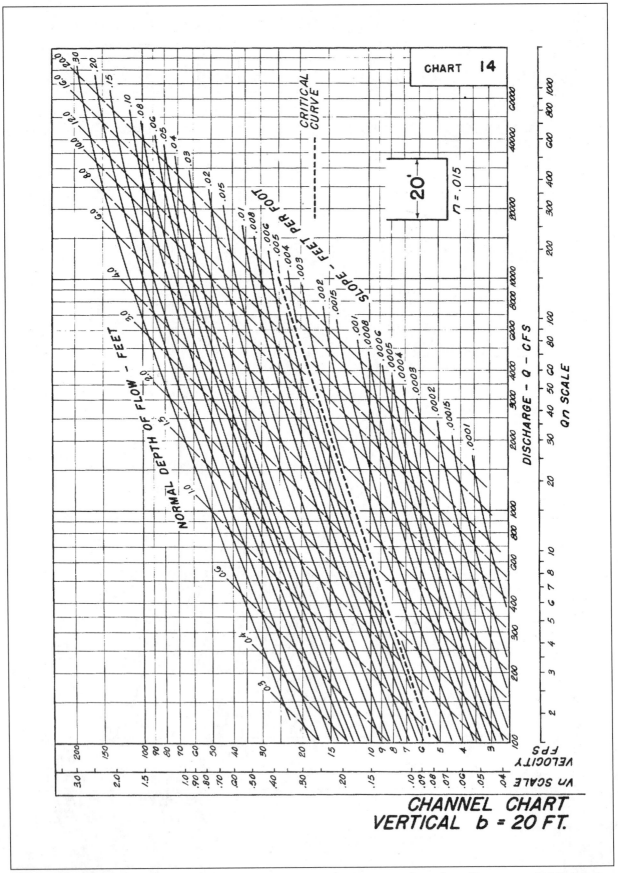

Figure A-3. (Chart 14) Channel charts for the solution of Manning's Equation. *(Courtesy of U.S. Dept. of Commerce, Bureau of Public Roads, Design Charts for Open-Channel Flow.)*

Figure A-3. (Chart 15)   Channel charts for the solution of Manning's Equation. *(Courtesy of U.S. Dept. of Commerce, Bureau of Public Roads, Design Charts for Open-Channel Flow.)*

Figure A-3. (Chart 16) Channel charts for the solution of Manning's Equation. *(Courtesy of U.S. Dept. of Commerce, Bureau of Public Roads, Design Charts for Open-Channel Flow.)*

Figure A-3. (Chart 17) Channel charts for the solution of Manning's Equation. *(Courtesy of U.S. Dept. of Commerce, Bureau of Public Roads, Design Charts for Open-Channel Flow.)*

Figure A-3. (Chart 18) Channel charts for the solution of Manning's Equation. *(Courtesy of U.S. Dept. of Commerce, Bureau of Public Roads, Design Charts for Open-Channel Flow.)*

CHART 19

CHANNEL CHART
2:1   b = 6 FT.

Figure A-3. (Chart 19)   Channel charts for the solution of Manning's Equation. *(Courtesy of U.S. Dept. of Commerce, Bureau of Public Roads, Design Charts for Open-Channel Flow.)*

Figure A-3. (Chart 21) Channel charts for the solution of Manning's Equation. *(Courtesy of U.S. Dept. of Commerce, Bureau of Public Roads, Design Charts for Open-Channel Flow.)*

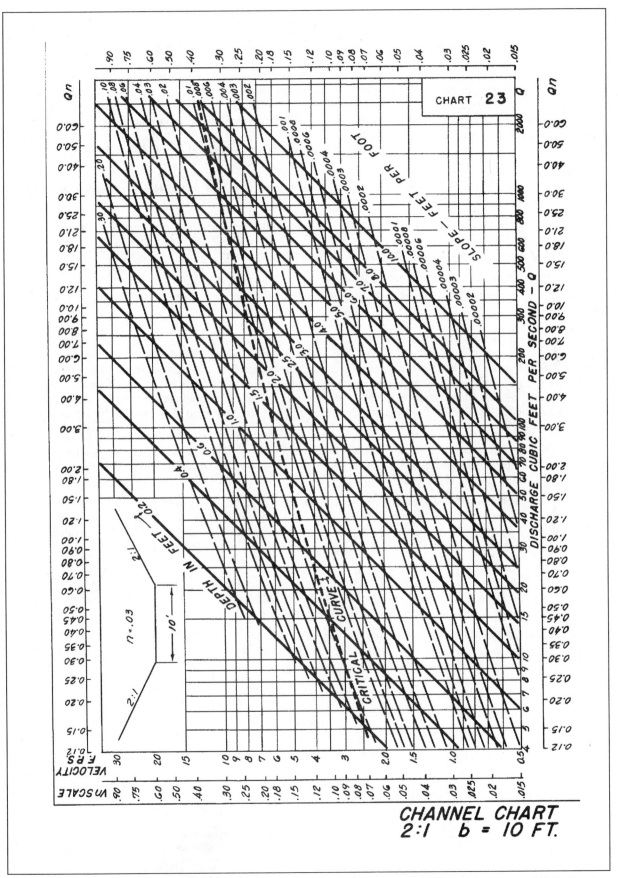

Figure A-3. (Chart 23)   Channel charts for the solution of Manning's Equation. (*Courtesy of U.S. Dept. of Commerce, Bureau of Public Roads, Design Charts for Open-Channel Flow.*)

A-4
Pipe Charts for the Solution of Manning's Equation

Charts 35 through 47 provide a direct graphical solution of Manning's Equation for circular pipes operating as an open channel (without pressure). The pipes are assumed to have uniform slope, cross-section, and roughness and are not affected by backwater.

The charts are calibrated for roughness values of 0.015, 0.012 and 0.024. However, as explained in Item 1 below, any n-value may be used.

You may use the charts to find discharge, velocity normal depth and critical depth. Following is a brief outline describing basic use of the charts:

1. **How to use the charts for any n-value**—If the roughness of your pipe is not one of the three listed above, follow this simple procedure:

   A. Adjust your known value of Q by multiplying by the ratio $\dfrac{n}{.015}$ to obtain an adjusted value, Q'.

   B. Enter Q' in the discharge scale for n = .015 and proceed as usual.

   If you are trying to determine discharge, follow this simple procedure:

   A. Using the usual procedure, drop a straight line to the discharge scale for n = .015.

   B. This value is actually Q', the adjusted discharge.

   C. Find the actual discharge by multiplying Q' by the ratio $\dfrac{.015}{n}$.

2. **How to find normal depth**—If you know discharge, roughness, and slope, find normal depth by entering the value of Q (or Q') and project a line straight up the graph until you intersect the appropriate slope line. (You may need to interpolate between slope lines.) At the point of intersection, read the normal depth using the series of diagonal lines labeled *NORMAL DEPTH OF FLOW IN PIPE—FEET*. (You may need to interpolate between lines.)

3. **How to find critical depth**—Critical depth is independent of roughness, so do not use the lower Q scales (n = 0.012, .024). If you know the discharge and slope, enter the value of Q and project straight up until you intersect the *CRITICAL* curve (dashed line). Using this point of intersection, read normal depth as in Item 2 above. The resulting value of normal depth is critical depth. (Remember that critical depth is a normal depth as well.)

4. **How to find velocity**—If you know discharge, roughness, and slope, find velocity by entering the value of Q (or Q') and project a line straight up until you intersect the appropriate slope line. (You may need to interpolate between slope lines.) Using this point of intersection, read the velocity using the series of curves running down and to the left labeled *VELOCITY—V—FPS*. (You may need to interpolate between lines.)

5. **How to find discharge**—If you know slope, normal depth, and roughness, find discharge by first locating the intersection of the appropriate slope and normal depth lines (interpolating if necessary). From the point of intersection, project a line straight down to the Q scale and read the value of discharge. (If necessary, adjust for a different n-value in accordance with Item 1 above.)

6. **How to find pipe capacity**—Pipe capacity actually is the same as finding discharge when normal depth equals pipe diameter. This value of normal depth is the last line in the series of normal depth lines and occurs at the end of the *hook* of the slope line as shown below.

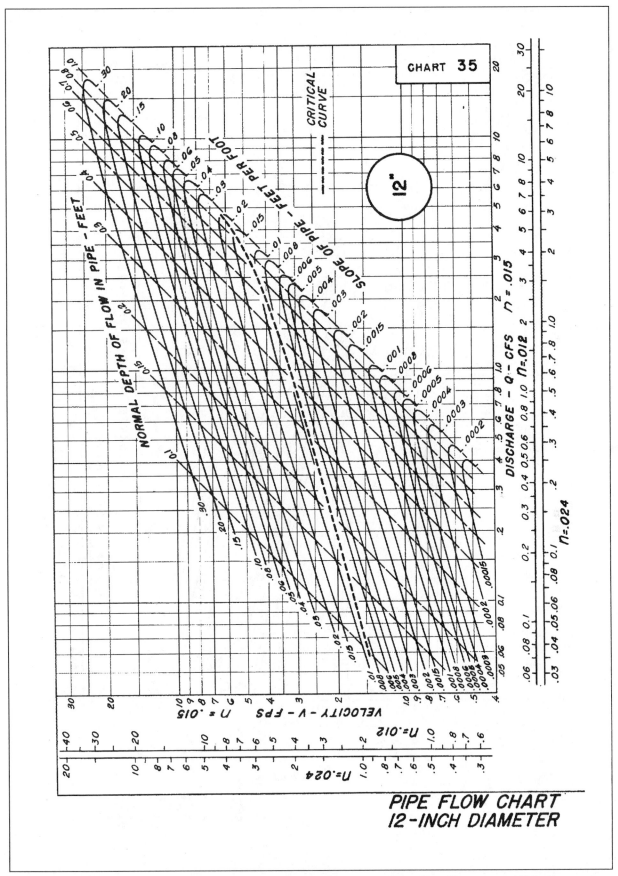

Figure A-4. (Chart 35)   Pipe charts for the solution of Manning's Equation. *(Courtesy of U.S. Dept. of Commerce, Bureau of Public Roads, Design Charts for Open-Channel Flow.)*

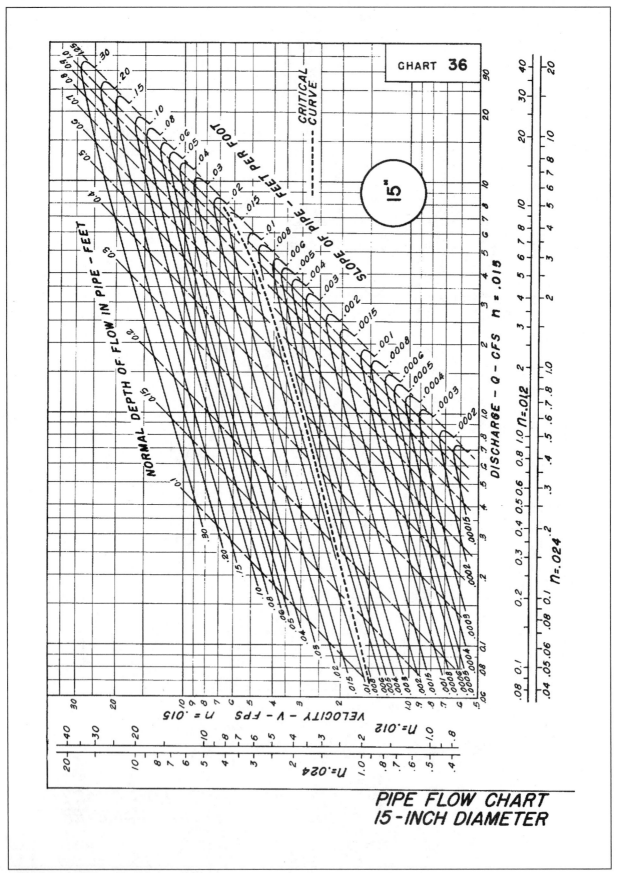

Figure A-4. (Chart 36) Pipe charts for the solution of Manning's Equation. *(Courtesy of U.S. Dept. of Commerce, Bureau of Public Roads, Design Charts for Open-Channel Flow.)*

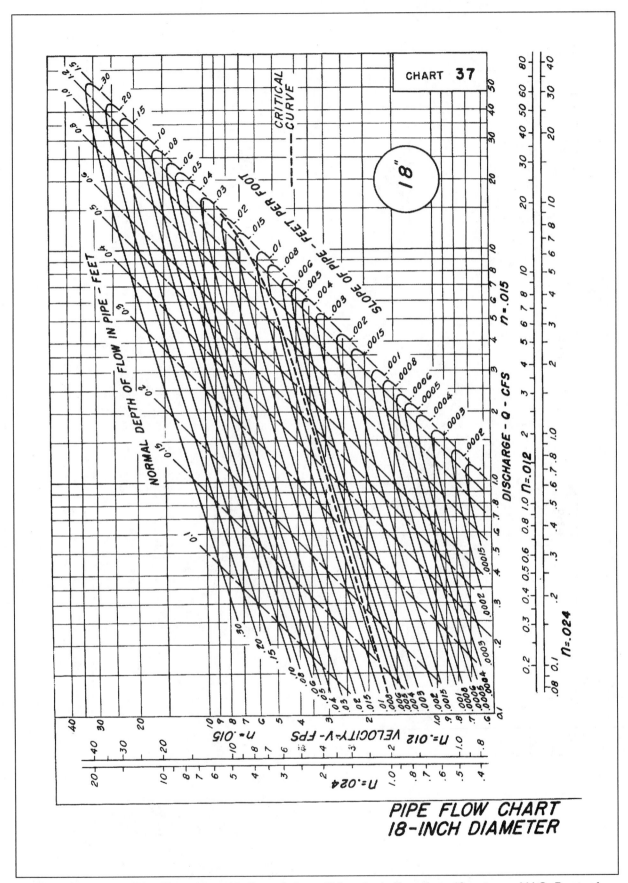

Figure A-4. (Chart 37)   Pipe charts for the solution of Manning's Equation. *(Courtesy of U.S. Dept. of Commerce, Bureau of Public Roads, Design Charts for Open-Channel Flow.)*

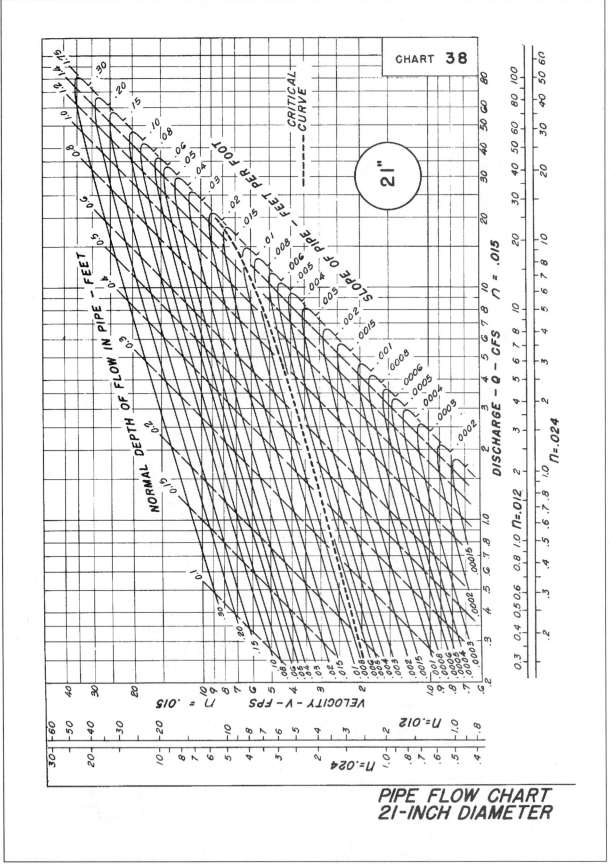

Figure A-4. (Chart 38)   Pipe charts for the solution of Manning's Equation. *(Courtesy of U.S. Dept. of Commerce, Bureau of Public Roads, Design Charts for Open-Channel Flow.)*

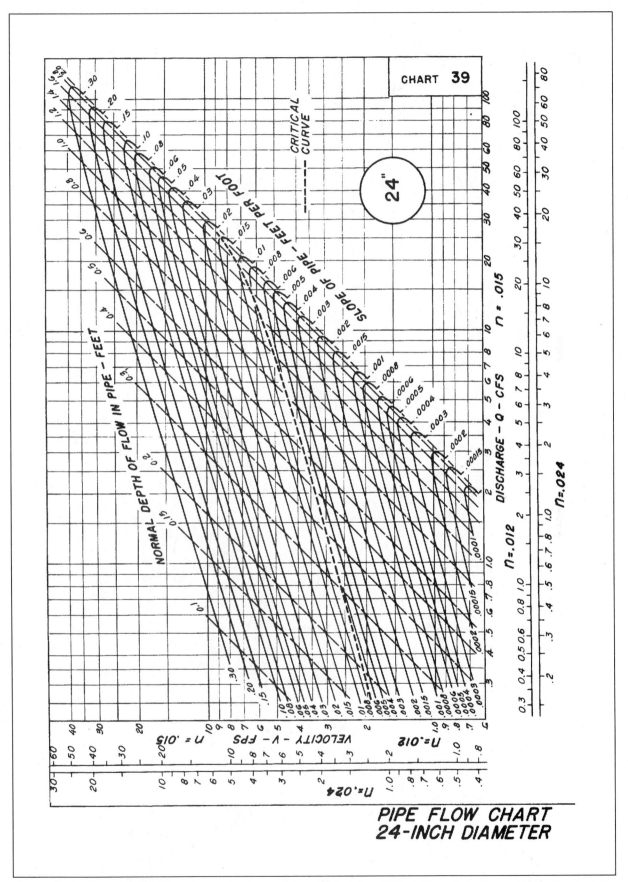

CHART 39

24"

PIPE FLOW CHART
24-INCH DIAMETER

Figure A-4. (Chart 39)   Pipe charts for the solution of Manning's Equation. *(Courtesy of U.S. Dept. of Commerce, Bureau of Public Roads, Design Charts for Open-Channel Flow.)*

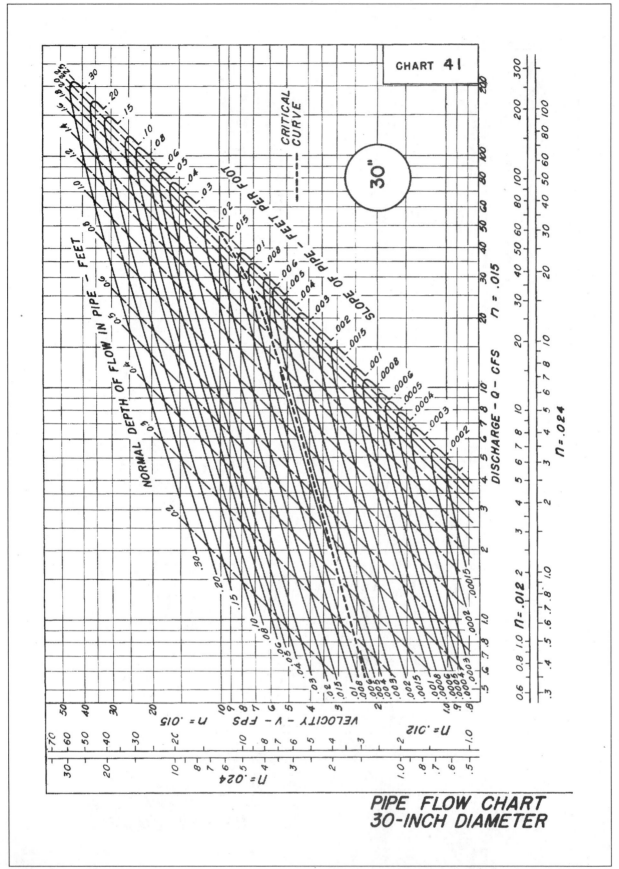

Figure A-4. (Chart 41)   Pipe charts for the solution of Manning's Equation. *(Courtesy of U.S. Dept. of Commerce, Bureau of Public Roads, Design Charts for Open-Channel Flow.)*

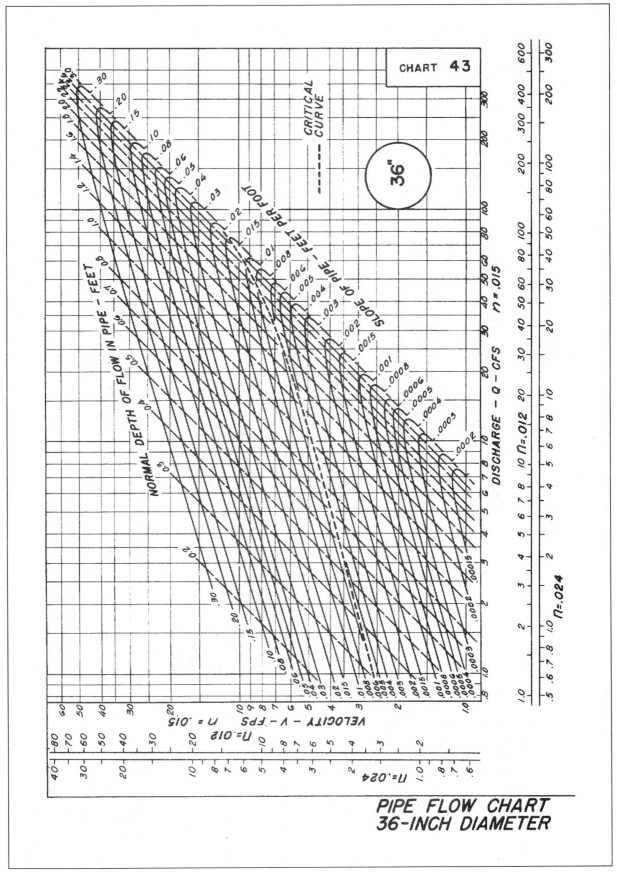

Figure A-4. (Chart 43)   Pipe charts for the solution of Manning's Equation. *(Courtesy of U.S. Dept. of Commerce, Bureau of Public Roads, Design Charts for Open-Channel Flow.)*

Figure A-4. (Chart 44) Pipe charts for the solution of Manning's Equation. *(Courtesy of U.S. Dept. of Commerce, Bureau of Public Roads, Design Charts for Open-Channel Flow.)*

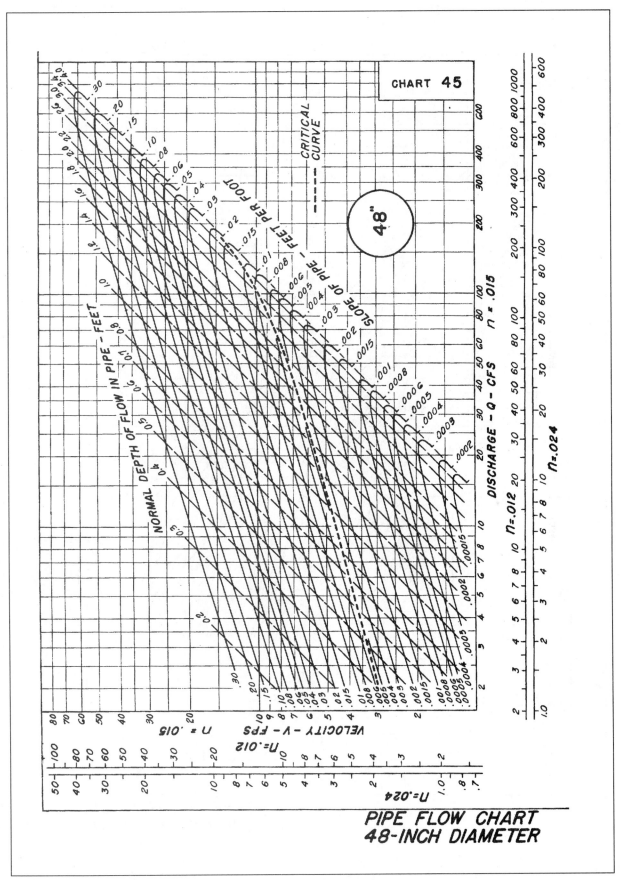

CHART 45

48"

PIPE FLOW CHART
48-INCH DIAMETER

Figure A-4. (Chart 45)   Pipe charts for the solution of Manning's Equation. *(Courtesy of U.S. Dept. of Commerce, Bureau of Public Roads, Design Charts for Open-Channel Flow.)*

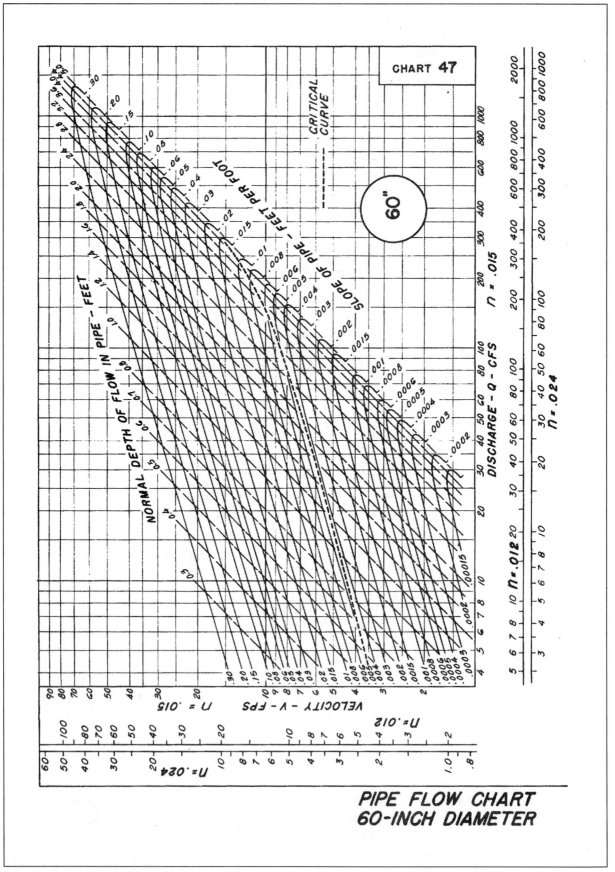

Figure A-4. (Chart 47) Pipe charts for the solution of Manning's Equation. *(Courtesy of U.S. Dept. of Commerce, Bureau of Public Roads, Design Charts for Open-Channel Flow.)*

A-5
Discharge Coefficients, c, for Broad Crested Weirs

Measured Head, H (ft)	Breadth of crest of weir (ft)										
	0.50	0.75	1.00	1.50	2.00	2.50	3.00	4.00	5.00	10.00	15.00
0.2	2.80	2.75	2.69	2.62	2.54	2.48	2.44	2.38	2.34	2.49	2.68
0.4	2.92	2.80	2.72	2.64	2.61	2.60	2.58	2.54	2.50	2.56	2.70
0.6	3.08	2.89	2.75	2.64	2.61	2.60	2.68	2.69	2.70	2.70	2.70
0.8	3.30	3.04	2.85	2.68	2.60	2.60	2.67	2.68	2.68	2.69	2.64
1.0	3.32	3.14	2.98	2.75	2.66	2.64	2.65	2.67	2.68	2.68	2.63
1.2	3.32	3.20	3.08	2.86	2.70	2.65	2.64	2.67	2.66	2.69	2.64
1.4	3.32	3.26	3.20	2.92	2.77	2.68	2.64	2.65	2.65	2.67	2.64
1.6	3.32	3.29	3.28	3.07	2.89	2.75	2.68	2.66	2.65	2.64	2.63
1.8	3.32	3.32	3.31	3.07	2.88	2.74	2.68	2.66	2.65	2.64	2.63
2.0	3.32	3.31	3.30	3.03	2.85	2.76	2.72	2.68	2.65	2.64	2.63
2.5	3.32	3.32	3.31	3.28	3.07	2.89	2.81	2.72	2.67	2.64	2.63
3.0	3.32	3.32	3.32	3.32	3.20	3.05	2.92	2.73	2.66	2.64	2.63
3.5	3.32	3.32	3.32	3.32	2.32	3.19	2.97	2.76	2.68	2.64	2.63
4.0	3.32	3.32	3.32	3.32	3.32	3.32	3.07	2.79	2.70	2.64	2.63
4.5	3.32	3.32	3.32	3.32	3.32	3.32	3.32	2.88	2.74	2.64	2.63
5.0	3.32	3.32	3.32	3.32	3.32	3.32	3.32	3.07	2.79	2.64	2.63
5.5	3.32	3.32	3.32	3.32	3.32	3.32	3.32	3.32	2.88	2.64	2.63

Figure A-5.   *(Courtesy of Brater & King, Handbook of Hydraulics, John Wiley & Sons, Inc.)*

B

DESIGN CHARTS FOR CULVERTS

B-1
Culverts with Inlet Control

Instructions for use:

1. How to find headwater depth, HW, given Q and size and type of culvert:
   a) Locate the height or diameter, D, of the culvert on the left-hand scale.
   b) Locate the discharge, Q, or Q/B for box culverts, on the center scale.
   c) Connect the two points in a and b above and extend the line to the right until it intersects the HW/D scale marked (1).
   d) If the HW/D scale marked (1) represents entrance type used, read HW/D on scale (1).
   e) If another of the three entrance types listed on the nomograph is used, extend the point of intersection made in c above horizontally to scale (2) or (3) and read HW/D.

2. How to find discharge, Q, per barrel, given HW and size and type of culvert:
   a) Compute HW/D for given conditions.
   b) Locate HW/D on scale (1), (2) or (3) for appropriate entrance type.
   c) If scale (2) or (3) is used, extend HW/D point horizontally to scale (1).
   d) Connect point on HW/D scale (1) and size of culvert, D, on left scale. Read Q or Q/B on the discharge scale.
   e) If Q/B is read in d above, multiply by B to find Q.

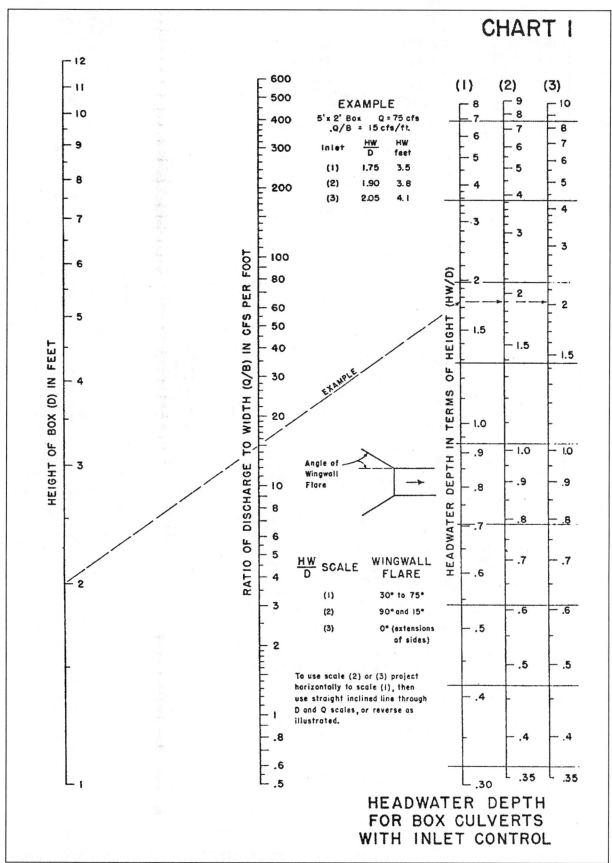

HEADWATER DEPTH
FOR BOX CULVERTS
WITH INLET CONTROL

Figure B-1. (Chart 1)   Culverts with inlet control. *(Courtesy of U.S. Dept. of Transportation, Federal Highway Administration, Hydraulic Charts for the Selection of Highway Culverts.)*

Figure B-1. (Chart 2)   Culverts with inlet control. *(Courtesy of U.S. Dept. of Transportation, Federal Highway Administration, Hydraulic Charts for the Selection of Highway Culverts.)*

B-2
Culverts with Outlet Control

Instructions for use:
1. How to find head, H, for a given culvert and discharge, Q:
   a) Locate appropriate nomograph for type of culvert selected. Find k_e for entrance type in Appendix B-3.
   b) Locate starting point on length scale. To locate the proper starting point on the length scales follow instructions below:
      (1) If the n value for the nomograph corresponds to that of the culvert being used, select the length curve for the proper k_e and locate the starting point at the given culvert length. If a k_e curve is not shown for the selected k_e, see (2) below. If the n value for the culvert selected differs from that of the nomograph, see (3) below.
      (2) for a k_e intermediate between the scales given, connect the given length on adjacent scales by a straight line and select a point on this line spaced between the two chart scales in proportion to the k_e values.
      (3) For a different roughness coefficient, n_1, than that of the chart n, use the length scales shown with an adjusted length L_1, calculated by the formula

$$L_1 = L(n_1/n)^2$$

   c) Locate the size of the culvert barrel on the culvert size scale (second scale from left).
   d) Connect the point on the size scale with the point on the length scale and note where the line passes through the *turning line*. (See Instruction 3 for size considerations for box culvert.)
   e) Locate the given discharge, Q, on the discharge scale.
   f) Connect the discharge point with the turning line point and extend the line to the head scale at the far right side of the nomograph. Read head, H, in feet on the head scale.
2. Values of n for commonly used culvert materials
   Concrete—Pipe 0.012
        Box 0.012
   Corrugated Metal

	Small Corrugations (2²/₃" x ¹/₂")	**Medium Corrugations** (3" x 1")	**Large Corrugations** (6" x 2")
Unpaved	0.024	0.027	varies*
25% paved	0.021	0.023	0.026
Fully paved	0.012	0.012	0.012

*n varies with diameter shown on charts. The various n values have been incorporated into the nomographs and no adjustment for culvert length is required.

3. To use the box culvert nomograph, Chart 8, for full-flow for other than square boxes:
   a) Compute cross-sectional area of the rectangular box.
   b) Locate area on the area scale.
   c) Connect point on area scale to point on length scale and note where the line passes through the turning line.
   d) Connect the discharge with the point on the turning line and extend the line to the head scale. Read head in feet on the head scale.

APPENDIX C—DESIGN CHARTS FOR RATIONAL METHOD

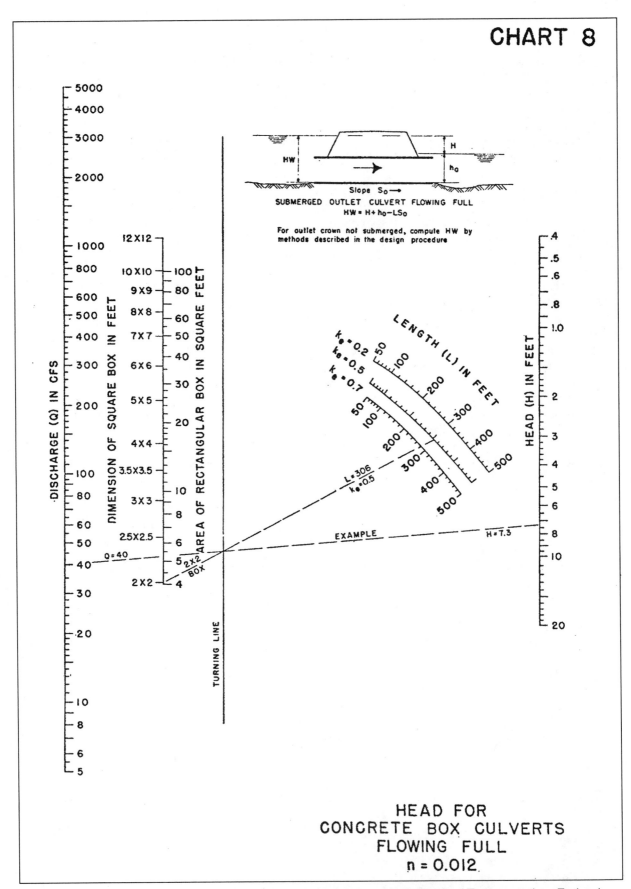

Figure B-2. (Chart 8) Culverts with outlet control. *(Courtesy of U.S. Dept. of Transportation, Federal Highway Administration, Hydraulic Charts for the Selection of Highway Culverts.)*

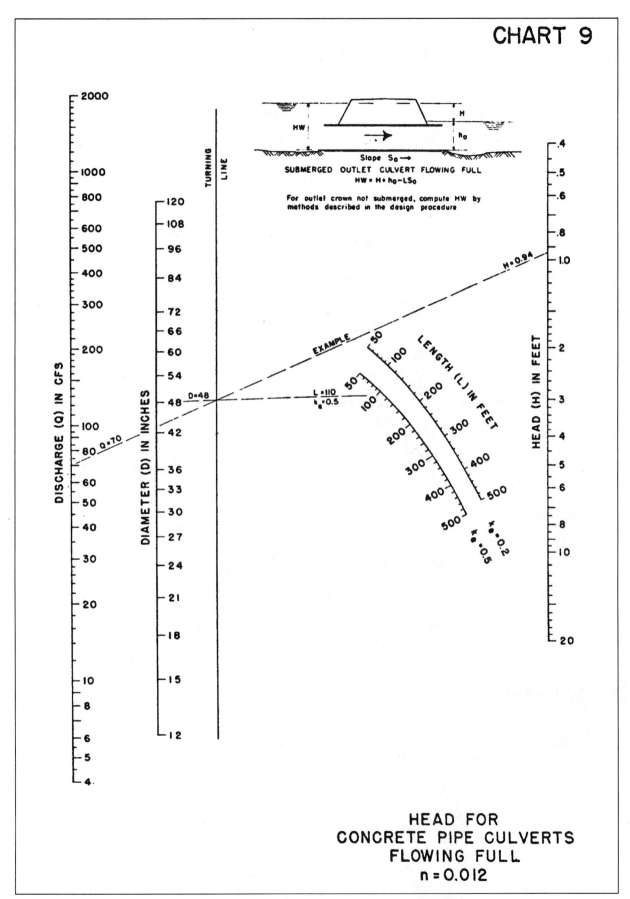

CHART 9

**HEAD FOR
CONCRETE PIPE CULVERTS
FLOWING FULL
n = 0.012**

Figure B-2. (Chart 9)   Culverts with outlet control. *(Courtesy of U.S. Dept. of Transportation, Federal Highway Administration, Hydraulic Charts for the Selection of Highway Culverts.)*

B-3
Entrance Loss Coefficients

Coefficient k_e to apply to velocity head $\dfrac{V^2}{2g}$ for determination of head loss at entrance to a structure, such as a culvert or conduit, operating full or partly full with *control at the outlet*.

$$\text{Entrance head loss } H_e = k_e \frac{V^2}{2g}$$

Type of Structure and Design of Entrance	Coefficient k_e
Pipe, Concrete	
Projecting from fill, socket end (groove-end)	0.2
Projecting from fill, sq. cut end	0.5
Headwall or headwall and wingwalls	
Socket end of pipe (groove-end)	0.2
Square-edge	0.5
Rounded (radius = $1/12D$)	0.2
Mitered to conform to fill slope	0.7
*End-Section conforming to fill slope	0.5
Pipe, or Pipe-Arch, Corrugated Metal	
Projecting from fill (no headwall)	0.9
Headwall or headwall and wingwalls	
Square-edge	0.5
Mitered to confrom to fill slope	0.7
*End-Section conforming to fill slope	0.5
Box, Reinforced Concrete	
Headwall parallel to embankment (no wingwalls)	
Square-edged on 3 edges	0.5
Rounded on 3 edges to radius of $1/12$ barrel dimension	0.2
Wingwalls at 30° to 75° to barrel	
Square-edged at crown	0.4
Crown edge rounded to radius of $1/12$ barrel dimension	0.2
Wingwalls at 10° to 25° to barrel	
Square-edged at crown	0.5
Wingwalls parallel (extension of sides)	
Square-edged at crown	0.7

*Note: "End Section conforming to fill slope," made of either metal or concrete, are the sections commonly available from manufacturers. From limited hydraulic tests they are equivalent in operation to a headwall in both *inlet* and *outlet* control. Some end sections, incorporating a *closed* taper in their design have a superior hydraulic performance. These latter sections can be designed using the information given for the bevelled inlet.

Figure B-3 Entrance loss coefficients. *(Courtesy of U.S. Dept. of Transportation, Federal Highway Administration, Hydraulic Charts for the Selection of Highway Culverts.)*

C

DESIGN CHARTS FOR RATIONAL METHOD

C-1
Values of c, Runoff Coefficient

Character of Surface	Runoff Coefficients
Pavement	
Asphalt and concrete	0.70 to 0.95
Brick	0.70 to 0.85
Roofs	0.75 to 0.95
Lawns, sandy soil	
Flat (2 percent)	0.05 to 0.10
Average (2 to 7 percent)	0.10 to 0.15
Steep (> 7 percent)	0.15 to 0.20
Lawns, heavy soil	
Flat (2 percent)	0.13 to 0.17
Average (2 to 7 percent)	0.18 to 0.22
Steep (> 7 percent)	0.25 to 0.35
Composite c-values:	
Business	
Downtown	0.70 to 0.95
Neighborhood	0.50 to 0.70
Residential	
Single Family	0.30 to 0.50
Multi-units, detached	0.40 to 0.60
Multi-units, attached	0.60 to 0.75
Residential (suburban)	0.25 to 0.40
Apartment	0.50 to 0.70
Industrial	
Light	0.50 to 0.80
Heavy	0.60 to 0.90
Parks, cemeteries	0.10 to 0.25
Playgrounds	0.20 to 0.35
Railroad yards	0.20 to 0.35
Unimproved	0.10 to 0.30

Note: The ranges of c values presented are typical for return periods of 2–10 years. Higher values are appropriate for larger design storms. Suggested multiplier factors for larger design storms are

Storm	Multiplier
25-year	1.15
50-year	1.20
100-year	1.25

Note: Adjusted c-value cannot exceed 1.00.

Figure C-1   Values of c, runoff coefficient. *(Courtesy of ASCE & Water Environmental Federation, Design and Construction of Urban Stormwater Management Systems.)*

C-2
Nomograph for Overland Flow Time

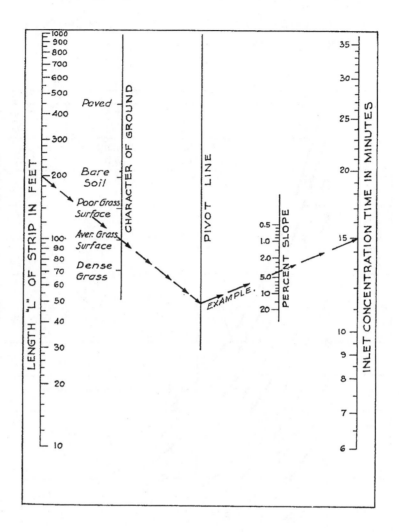

Figure C-2   Nomograph for overland flow time. *(Courtesy of E. Seeley, Data Book for Civil Engineers, Vol. 1, John Wiley & Sons, Inc.)*

C-3
Intensity-Duration-Frequency (I-D-F) Curves

Example: For a selected 10-year return period, P$_1$ = 2.0 inches. T$_C$ is calculated as 20 minutes. Therefore, (i) = 4.25 in/hr.

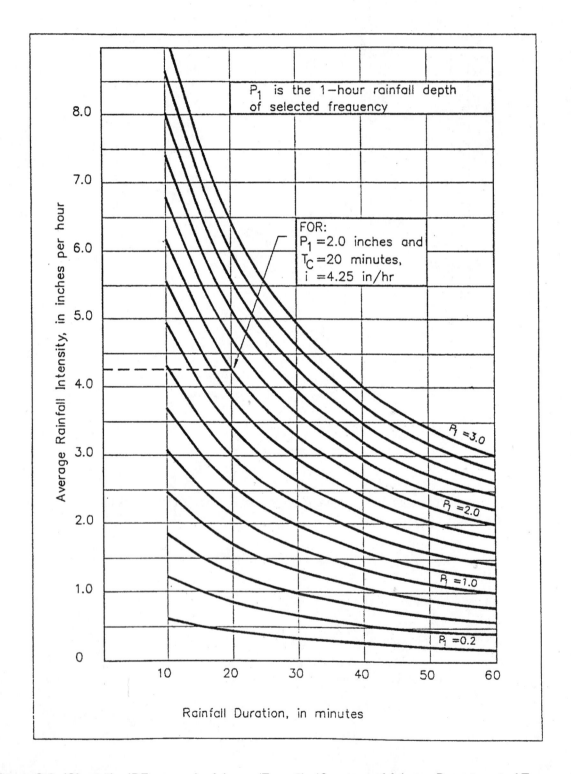

Figure C-3. (Chart 1) IDF curves for Arizona (Zone 6). *(Courtesy of Arizona Department of Transportation.)*

Figure C-3. (Chart 2) IDF curves for Orange County, California. *(Courtesy of California Department of Transportation.)*

Figure C-3. (Chart 3) IDF curves for Atlanta, Georgia. *(Courtesy of Georgia Department of Transportation.)*

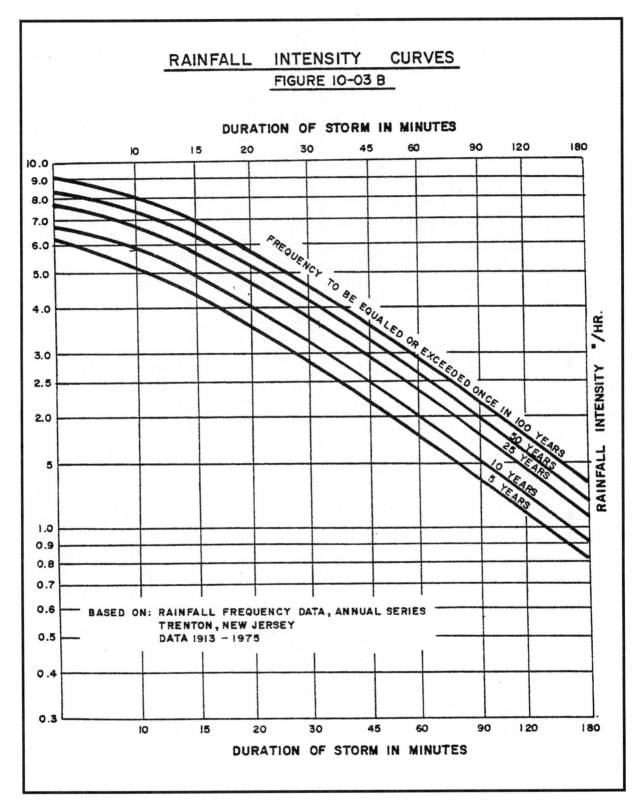

Figure C-3. (Chart 4) IDF curves for New Jersey. *(Courtesy of New Jersey Department of Transportation.)*

Figure C-3. (Chart 5) IDF curves for Pennsylvania (Region 1). *(Courtesy of Pennsylvania Department of Transportation.)*

D

DESIGN CHARTS FOR SCS METHOD

D-1
Runoff Curve Numbers

Cover description		Curve numbers for hydrologic soil group—			
Cover type and hydrologic condition	Average percent impervious area[2]	A	B	C	D
Fully developed urban areas (vegetation established)					
Open space (lawns, parks, golf courses, cemeteries, etc.)[3]:					
Poor condition (grass cover < 50%)		68	79	86	89
Fair condition (grass cover 50% to 75%)...........		49	69	79	84
Good condition (grass cover > 75%)		39	61	74	80
Impervious areas:					
Paved parking lots, roofs, driveways, etc. (excluding right-of-way).		98	98	98	98
Streets and roads:					
Paved; curbs and storm sewers (excluding right-of-way)...................................		98	98	98	98
Paved; open ditches (including right-of-way)		83	89	92	93
Gravel (including right-of-way)		76	85	89	91
Dirt (including right-of-way)		72	82	87	89
Western desert urban areas:					
Natural desert landscaping (pervious areas only)[4]...		63	77	85	88
Artificial desert landscaping (impervious weed barrier, desert shrub with 1- to 2-inch sand or gravel mulch and basin borders).		96	96	96	96
Urban districts:					
Commercial and business.........................	85	89	92	94	95
Industrial......................................	72	81	88	91	93
Residential districts by average lot size:					
1/8 acre or less (town houses).....................	65	77	85	90	92
1/4 acre	38	61	75	83	87
1/3 acre	30	57	72	81	86
1/2 acre	25	54	70	80	85
1 acre	20	51	68	79	84
2 acres	12	46	65	77	82
Developing urban areas					
Newly graded areas (pervious areas only, no vegetation)[5]		77	86	91	94
Idle lands (CN's are determined using cover types similar to those in table 2-2c).					

[1]Average runoff condition, and $I_a = 0.2S$.

[2]The average percent impervious area shown was used to develop the composite CN's. Other assumptions are as follows: impervious areas are directly connected to the drainage system, impervious areas have a CN of 98, and pervious areas are considered equivalent to open space in good hydrologic condition. CN's for other combinations of conditions may be computed using figure 2-3 or 2-4.

[3]CN's shown are equivalent to those of pasture. Composite CN's may be computed for other combinations of open space cover type.

[4]Composite CN's for natural desert landscaping should be computed using figures 2-3 or 2-4 based on the impervious area percentage (CN = 98) and the pervious area CN. The pervious area CN's are assumed equivalent to desert shrub in poor hydrologic condition.

[5]Composite CN's to use for the design of temporary measures during grading and construction should be computed using figure 2-3 or 2-4, based on the degree of development (impervious area percentage) and the CN's for the newly graded pervious areas.

Figure D-1. (Chart 1)   Runoff curve numbers. *(Courtesy of Soil Conservation Service, Technical Release 55.)*

Cover description		Curve numbers for hydrologic soil group—			
Cover type	Hydrologic condition	A	B	C	D
Pasture, grassland, or range—continuous forage for grazing.[2]	Poor	68	79	86	89
	Fair	49	69	79	84
	Good	39	61	74	80
Meadow—continuous grass, protected from grazing and generally mowed for hay.	—	30	58	71	78
Brush—brush-weed-grass mixture with brush the major element.[3]	Poor	48	67	77	83
	Fair	35	56	70	77
	Good	[4]30	48	65	73
Woods—grass combination (orchard or tree farm).[5]	Poor	57	73	82	86
	Fair	43	65	76	82
	Good	32	58	72	79
Woods.[6]	Poor	45	66	77	83
	Fair	36	60	73	79
	Good	[4]30	55	70	77
Farmsteads—buildings, lanes, driveways, and surrounding lots.	—	59	74	82	86

[1]Average runoff condition, and $I_a = 0.2S$.

[2]*Poor:*   <50% ground cover or heavily grazed with no mulch.
Fair: 50 to 75% ground cover and not heavily grazed.
Good: >75% ground cover and lightly or only occasionally grazed.

[3]*Poor:*   <50% ground cover.
Fair: 50 to 75% ground cover.
Good: >75% ground cover.

[4]Actual curve number is less than 30; use CN = 30 for runoff computations.

[5]CN's shown were computed for areas with 50% woods and 50% grass (pasture) cover. Other combinations of conditions may be computed from the CN's for woods and pasture.

[6]*Poor:*   Forest litter, small trees, and brush are destroyed by heavy grazing or regular burning.
Fair: Woods are grazed but not burned, and some forest litter covers the soil.
Good: Woods are protected from grazing, and litter and brush adequately cover the soil.

Figure D-1. (Chart 2)   Runoff curve numbers. *(Courtesy of Soil Conservation Service, Technical Release 55.)*

D-2
Hydrologic Soil Groups for Selected Soil Types

Aabab	D		Lamington	D
Alluvial land	C		Lansing	B
Amwell	C		Lehigh	C
Avoca	B		Linkville	B
Bartley	C		Louisa	B
Belmill	B		Lynchburg	C
Bigbee	A		Madden	C
Boonton	C		Madrid	B
Broadwell	B		Maplecrest	B
Bucks	B		McDaniel	B
Califon	C		Menlo	D
Chancelor	C		Minoa	C
Chesire	B		Mirkwood	D
Conic	C		Modesto	C
Continental	C		Moundhaven	A
Custer	D		Muskingum	C
Delaney	A		Narragansett	B
Dickerson	D		Netcong	B
Dundee	C		Nobscot	A
Eastland	B		Nuff	C
Edneyville	B		Okeetee	D
Exeter	C		Outlet	C
Fairchild	C		Oxbow	C
Freedom	C		Parker	B
Fulerton	B		Penn	C
Georgetown	D		Pompton	B
Gleason	B		Preakness	C
Goosebury	B		Punsit	C
Gravel pits	A		Rayburn	D
Grigston	B		Reading	B
Haledon	C		Riverhead	B
Hamel	C		Rodessa	D
Haven	B		Santa Fe	D
Herty	D		Scarboro	D
Hotsprings	B		Shellbluff	B
Hungry	C		Snelling	B
Idlewild	D		Summers	B
Isabella	B		Tarrytown	C
Jardin	D		Terril	B
Jigsaw	C		Troutdale	C
Jonesville	B		Urban land	(varies)
Joplin	C		Utica	B
Kaplan	D		Vercliff	C
Kehar	D		Vinsad	C
Kilkenny	B		Washingston	B
Kirkville	C		Whitecross	D
Kreamer	C		Yardley	C
Lambert	B		Zynbar	B

Figure D-2   Hydrologic soil groups for selected soil types. *(Courtesy of Soil Conservation Service, Technical Release 55.)*

D-3
24-hour Rainfall Amounts by Return Period

The following maps are reproduced from the *Rainfall Atlas of the United States,* Technical Paper No. 40 by the U.S. Weather Bureau. The maps show isopluvial lines (lines of equal rainfall depth) for 100, 50, 25, 10, 5, and 2-year, 24-hour rainfalls throughout the contiguous 48 states.

Also included are selected maps from the National Oceanic and Atmospheric Administration (NOAA) Atlas 2 for Arizona and Southern California for the 100-year, 24-hour rainfall.

Figure D-3. (Map 1)   Continental U.S., 100-year, 24-hour rainfall (inches). *(Courtesy of U.S. Weather Bureau, Rainfall Atlas of the United States.)*

Figure D-3. (Map 2)   Continental U.S., 50-year, 24-hour rainfall (inches). *(Courtesy of U.S. Weather Bureau, Rainfall Atlas of the United States.)*

Figure D-3. (Map 3) Continental U.S., 25-year, 24-hour rainfall (inches). *(Courtesy of U.S. Weather Bureau, Rainfall Atlas of the United States.)*

Figure D-3. (Map 4)   Continental U.S., 10-year, 24-hour rainfall (inches). *(Courtesy of U.S. Weather Bureau, Rainfall Atlas of the United States.)*

Figure D-3. (Map 5) Continental U.S., 5-year, 24-hour rainfall (inches). *(Courtesy of U.S. Weather Bureau, Rainfall Atlas of the United States.)*

Figure D-3. (Map 6)   Continental U.S., 2-year, 24-hour rainfall (inches). *(Courtesy of U.S. Weather Bureau, Rainfall Atlas of the United States.)*

Figure 29
ISOPLUVIALS OF 100-YR 24-HR PRECIPITATION
IN TENTHS OF AN INCH

ARIZONA

10 0 10 20 30 40
MILES

NOAA ATLAS 2, Volume VIII
Prepared by U.S. Department of Commerce
National Oceanic and Atmospheric Administration
National Weather Service, Office of Hydrology
Prepared for U.S. Department of Agriculture,
Soil Conservation Service, Engineering Division

Figure D-3. (Map 7) Arizona, 100-year, 24-hour rainfall (tenths of an inch). *(Courtesy of NOAA Atlas 2, Volume VIII.)*

Figure D-3. (Map 8)   Southern California, 100-year, 24-hour rainfall (tenths of an inch). *(Courtesy of NOAA Atlas 2, Volume XI.)*

D-4
Approximate Geographical Boundaries for SCS Rainfall Distributions

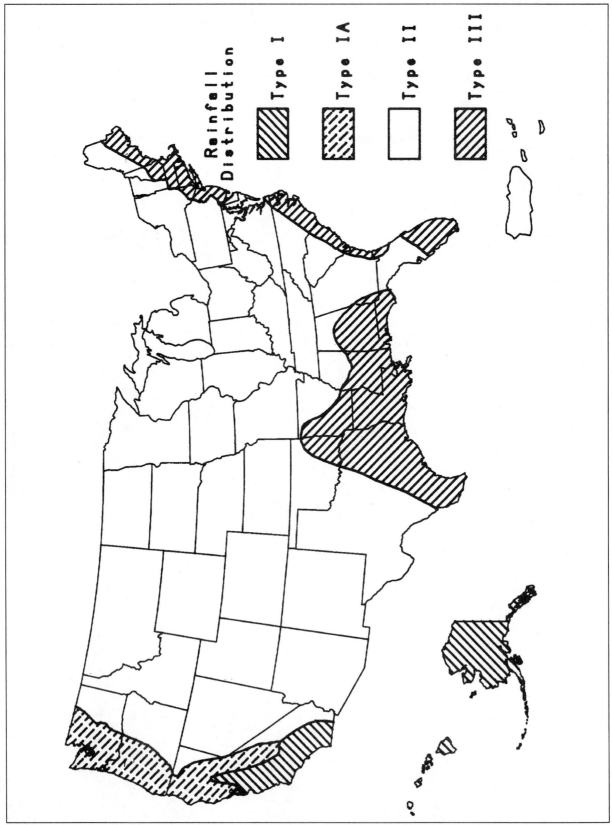

Figure D-4 Approximate geographical boundaries for SCS rainfall distributions. *(Courtesy of Soil Conservation Service, Technical Release 55.)*

D-5
Unit Peak Discharge (q$_u$)

Figure D-5. (Chart 1) Unit peak discharge (q$_u$) for SCS type I rainfall distribution. *(Courtesy of Soil Conservation Service, Technical Release 55.)*

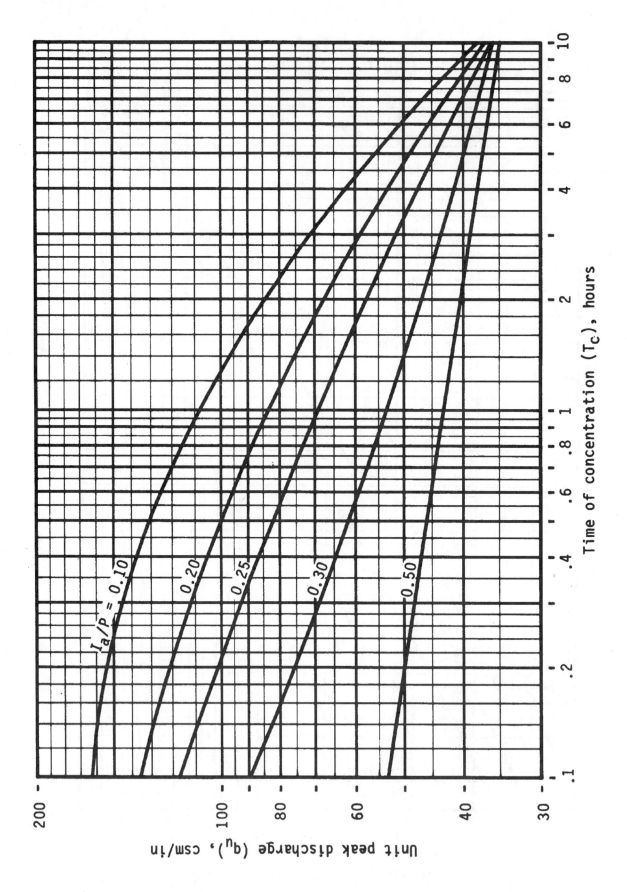

Figure D-5. (Chart 2) Unit peak discharge (q$_u$) for SCS type IA rainfall distribution. *(Courtesy of Soil Conservation Service, Technical Release 55.)*

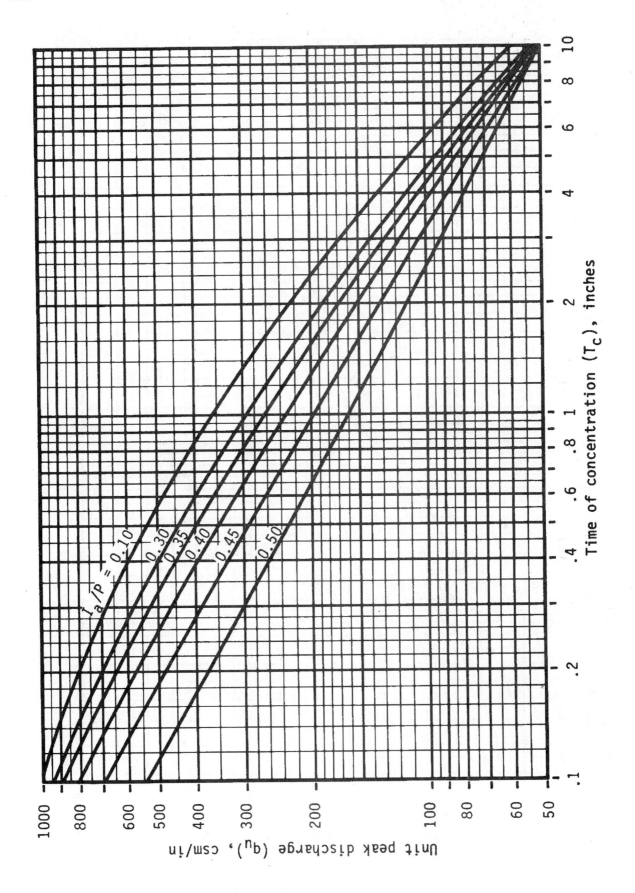

Figure D-5. (Chart 3)   Unit peak discharge (q$_u$) for SCS type II rainfall distribution. *(Courtesy of Soil Conservation Service, Technical Release 55.)*

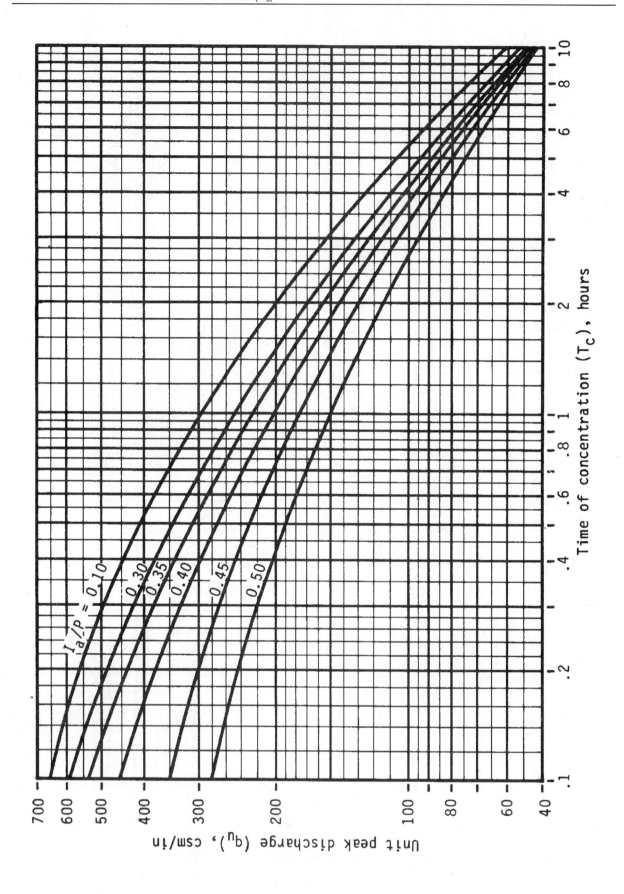

Figure D-5. (Chart 4) Unit peak discharge (q$_u$) for SCS type III rainfall distribution. *(Courtesy of Soil Conservation Service, Technical Release 55.)*

D-6
Tabular Hydrograph Unit Discharges (csm/in) for Type II Rainfall Distributions

HYDROGRAPH TIME (HOURS) — Rainfall Type = II — $t_c = 0.4$ hr

IA/P = 0.10

TRVL TIME (HR)	11.0	11.3	11.6	11.9	12.0	12.1	12.2	12.3	12.4	12.5	12.6	12.7	12.8	13.0	13.2	13.4	13.6	13.8	14.0	14.3	14.6	15.0	15.5	16.0	16.5	17.0	17.5	18.0	19.0	20.0	22.0	26.0
0.0	18	25	36	77	141	271	468	592	574	431	298	216	163	104	77	63	55	49	44	38	34	31	28	25	22	21	20	18	16	14	12	0
.10	18	24	34	67	116	219	385	557	523	357	196	196	119	84	67	57	51	56	51	39	35	32	29	26	23	21	20	18	16	14	12	0
.20	15	20	28	44	59	116	179	316	454	523	489	401	309	178	112	81	65	56	49	44	37	33	30	26	23	21	20	19	17	14	12	0
.30	15	20	27	41	53	82	147	260	389	478	486	429	349	210	129	89	69	58	51	43	38	33	30	27	24	21	20	19	17	14	12	1
.40	13	17	23	33	38	48	71	121	214	331	429	467	442	308	189	120	85	66	56	47	41	35	31	28	24	22	20	19	17	15	13	0
.50	12	16	22	31	36	44	62	102	176	279	379	438	440	339	218	137	94	71	59	49	42	35	31	28	25	22	21	19	17	15	14	0
.75	11	13	17	24	26	30	35	45	65	106	170	251	326	393	341	245	164	112	81	59	48	39	33	30	26	23	21	20	18	15	15	0
1.0	8	10	13	17	19	21	24	27	31	37	50	75	118	251	360	376	292	205	138	83	60	45	36	32	28	25	22	21	18	16	16	1
1.5	6	7	9	12	13	14	15	17	19	21	23	26	31	56	121	224	311	333	293	192	115	66	45	36	31	28	25	22	19	17	13	4
2.0	4	5	6	8	9	9	10	10	11	12	14	15	16	20	27	43	85	159	243	306	264	154	74	47	37	32	28	25	21	18	15	9
2.5	2	3	4	5	6	6	7	7	8	9	11	10	11	13	16	20	27	46	85	184	285	262	147	74	47	37	32	28	22	19	16	11
3.0	1	2	2	3	4	4	4	5	5	6	6	7	7	8	10	12	14	17	23	47	109	227	268	160	83	50	38	32	25	21	16	11

IA/P = 0.30

TRVL TIME (HR)	11.0	11.3	11.6	11.9	12.0	12.1	12.2	12.3	12.4	12.5	12.6	12.7	12.8	13.0	13.2	13.4	13.6	13.8	14.0	14.3	14.6	15.0	15.5	16.0	16.5	17.0	17.5	18.0	19.0	20.0	22.0	26.0
0.0	4	26	113	296	480	495	413	306	234	186	127	100	84	67	61	54	49	45	41	37	33	31	29	28	25	22	21	20	19	19	19	0
.10	0	18	81	224	395	462	430	347	272	172	96	82	74	66	61	57	51	46	42	38	35	32	30	28	25	23	22	20	19	18	17	0
.20	0	2	13	59	169	224	305	373	414	305	196	134	103	75	67	59	52	47	43	39	34	32	30	26	25	23	22	20	19	17	15	0
.30	0	0	1	9	142	306	383	403	361	255	181	127	99	83	73	63	55	48	44	40	36	32	29	27	24	23	21	20	19	17	11	0
.40	0	0	0	1	30	94	202	298	372	379	308	203	141	100	84	67	61	56	49	47	41	38	34	31	29	28	26	24	23	19	11	0
.50	0	0	0	0	4	21	70	158	258	334	347	364	270	172	96	82	74	66	61	57	51	46	43	40	36	33	30	28	25	19	17	0
.75	0	0	0	0	1	8	30	76	145	219	305	321	305	393	341	196	134	130	102	78	65	55	47	43	39	34	31	29	27	23	17	0
1.0	0	0	0	0	0	0	2	15	42	127	255	403	383	308	360	272	209	154	103	79	63	55	48	45	41	37	33	31	24	21	18	1
1.5	0	0	0	0	0	0	0	0	10	51	136	226	263	274	226	195	131	85	62	51	45	41	37	33	31	29	26	25	20	15	0	6
2.0	0	0	0	0	0	0	0	0	0	0	0	18	31	86	162	239	252	86	31	3	0	0	0	0	0	0	0	0	0	0	0	15
2.5	0	0	0	0	0	0	0	0	0	0	0	1	13	31	112	202	235	155	92	33	0	0	0	0	0	0	0	0	0	0	0	18
3.0	0	0	0	0	0	0	0	0	0	0	0	0	1	21	76	154	221	182	76	21	0	0	0	0	0	0	0	0	0	0	0	18

IA/P = 0.50

TRVL TIME (HR)	11.0	11.3	11.6	11.9	12.0	12.1	12.2	12.3	12.4	12.5	12.6	12.7	12.8	13.0	13.2	13.4	13.6	13.8	14.0	14.3	14.6	15.0	15.5	16.0	16.5	17.0	17.5	18.0	19.0	20.0	22.0	26.0
0.0	7	59	168	245	257	213	186	163	128	109	96	88	81	75	67	62	58	54	50	45	43	41	39	35	31	29	28	28	28	28	28	1
.10	0	5	41	125	205	240	222	198	163	135	123	106	94	86	79	71	64	60	56	51	46	45	43	37	35	33	31	29	34	32	31	0
.20	0	0	3	28	93	168	216	220	205	168	131	110	97	88	81	72	65	56	51	46	43	42	40	36	35	34	32	30	38	34	32	0
.30	0	0	0	2	20	69	135	189	209	169	135	107	95	86	77	69	63	58	53	48	44	42	41	37	36	35	33	31	42	35	33	1
.40	0	1	14	50	106	161	193	202	98	70	62	58	53	48	44	41	37	33	31	29	26	24	23	21	20	19	18	17	37	40	45	7
.50	0	0	9	37	83	135	174	194	102	71	63	60	54	48	45	43	39	34	32	31	28	25	23	22	21	20	18	18	39	42	48	20
.75	0	0	0	0	4	21	53	107	146	124	86	68	60	56	51	47	43	40	37	35	34	33	30	28	27	26	25	24	40	45	53	26
1.0	0	0	0	0	0	0	1	21	78	141	146	173	125	101	86	77	69	63	57	53	48	45	42	39	36	35	31	29	42	48	57	27
1.5	0	0	0	0	0	0	0	0	14	50	72	89	113	159	153	139	113	98	75	67	62	58	53	50	48	48	41	39	44	40	29	7
2.0	0	1	9	37	83	125	16	3	9	37	93	125	150	138	86	51	60	56	54	51	48	45	43	40	37	35	33	31	48	42	31	20
2.5	0	0	0	0	0	0	0	0	9	41	57	93	121	143	177	124	65	17	0	0	0	0	0	0	0	0	0	0	53	45	32	26
3.0	0	0	0	0	0	0	0	0	0	2	20	63	107	146	138	101	40	11	0	0	0	0	0	0	0	0	0	0	57	48	34	27

Figure D-6. (Chart 1) Tabular hydrograph unit discharges (csm/in) for type II rainfall distribution (t_c = 0.4 hour). *(Courtesy of Soil Conservation Service, Technical Release 55.)*

HYDROGRAPH TIME (HOURS) — RAINFALL TYPE = II — TC = 0.5 HR

IA/P = 0.10

TRVL TIME (HR)	11.0	11.3	11.6	11.9	12.0	12.1	12.2	12.3	12.4	12.5	12.6	12.7	12.8	13.0	13.2	13.4	13.6	13.8	14.0	14.3	14.6	15.0	15.5	16.0	16.5	17.0	17.5	18.0	19.0	20.0	22.0	26.0
0.0	17	23	32	57	94	170	308	467	529	507	402	297	226	140	96	74	61	53	47	41	36	32	29	26	23	21	20	19	16	14	12	0
.10	16	22	30	51	80	140	252	395	484	499	434	343	265	162	108	80	65	55	49	42	36	33	29	26	23	21	20	19	16	14	12	0
.20	14	19	25	38	47	69	116	207	332	434	477	449	378	238	149	101	77	62	53	45	39	34	30	27	24	22	20	17	14	14	12	1
.30	13	18	24	35	43	60	97	170	278	382	446	448	401	270	171	114	83	66	56	46	40	34	31	27	24	22	20	17	15	12	12	1
.40	12	15	21	29	40	53	83	141	233	332	408	434	361	243	157	107	79	64	51	43	36	32	28	26	23	21	17	17	12	12	11	1
.50	11	15	20	28	37	48	71	118	194	286	367	412	378	271	178	119	86	68	53	44	37	32	29	25	23	20	17	15	12	12	11	1
.75	9	11	14	19	24	27	31	49	74	118	169	182	319	328	244	136	66	24	18	16	31	35	31	31	29	33	29	29	20	17	16	1
1.0	7	9	11	16	21	19	21	32	40	55	83	188	309	359	245	172	20	13	11	102	68	49	38	32	29	26	23	21	17	16	11	1
1.5	5	7	8	11	12	13	14	15	17	19	21	23	27	43	89	140	225	322	309	202	88	52	39	33	29	23	20	17	13	11	11	5
2.0	3	4	4	7	8	8	9	10	11	11	14	15	15	18	23	35	66	123	202	297	181	87	52	39	31	26	21	19	14	11	11	10
2.5	1	2	2	3	4	4	6	7	7	7	7	8	10	12	15	18	24	36	66	150	278	171	87	52	31	29	23	20	15	11	11	11
3.0	1	1	1	2	3	3	4	4	5	6	6	8	11	13	16	20	86	198	263	182	96	56	40	33	26	21	16	11	11	11

IA/P = 0.30

TRVL TIME (HR)	11.0	11.3	11.6	11.9	12.0	12.1	12.2	12.3	12.4	12.5	12.6	12.7	12.8	13.0	13.2	13.4	13.6	13.8	14.0	14.3	14.6	15.0	15.5	16.0	16.5	17.0	17.5	18.0	19.0	20.0	22.0	26.0
0.0	0	0	9	53	157	314	433	439	379	299	237	159	118	89	118	159	218	268	309	330	391	388	341	372	316	223	156	117	94	...	...	...
.10	0	0	7	37	117	248	372	416	391	330	218	150	113	95	81	71	65	60	56	53	50	47	43	39	35	32	30	29	26	25	22	19
.20	0	0	4	26	87	194	313	382	388	349	244	167	122	97	82	72	66	62	58	54	48	43	39	39	35	32	30	29	26	23	19
.30	0	0	0	19	64	151	259	341	372	341	316	223	156	117	94	80	80	67	58	50	45	41	36	33	31	29	26	23	19

IA/P = 0.50

TRVL TIME (HR)	11.0	11.3	11.6	11.9	12.0	12.1	12.2	12.3	12.4	12.5	12.6	12.7	12.8	13.0	13.2	13.4	13.6	13.8	14.0	14.3	14.6	15.0	15.5	16.0	16.5	17.0	17.5	18.0	19.0	20.0	22.0	26.0
0.0	0	0	0	2	26	89	170	200	229	217	179	144	119	104	93	85	78	70	64	59	55	51	46	43	41	40	36	32	28
.10	0	0	0	0	18	65	135	205	190	170	137	115	101	91	83	76	70	67	61	56	52	47	44	42	40	38	32	28
.20	0	0	0	0	12	47	106	162	198	178	145	121	105	94	85	76	72	68	61	57	52	48	44	40	38	34	28
.30	0	0	0	0	8	34	82	135	177	194	168	139	117	102	92	80	71	65	58	54	49	45	41	37	35	33	28

Figure D-6. (Chart 2) Tabular hydrograph unit discharges (csm/in) for type II rainfall distribution (t_c = 0.5 hour). *(Courtesy of Soil Conservation Service, Technical Release 55.)*

RAINFALL TYPE = II

* * * TC = 1.0 HR * * *

HYDROGRAPH TIME (HOURS)

IA/P = 0.10

TRVL TIME (HR)	11.0	11.3	11.6	11.9	12.0	12.1	12.2	12.3	12.4	12.5	12.6	12.7	12.8	13.0	13.2	13.4	13.6	13.8	14.0	14.3	14.6	15.0	15.5	16.0	16.5	17.0	17.5	18.0	19.0	20.0	22.0	26.0
0.0	11	15	20	29	35	47	72	112	168	231	289	329	357	313	239	175	133	103	83	63	50	40	33	29	26	23	21	20	17	15	12	0
.10	10	13	17	24	27	33	42	62	95	144	202	260	306	340	293	222	165	126	98	72	56	43	35	30	27	24	22	20	18	15	12	0
.20	10	13	17	23	26	30	38	54	82	123	176	232	281	332	303	238	179	136	105	76	59	45	35	30	27	24	22	20	18	16	12	1
.30	9	12	16	22	24	28	35	48	70	105	152	205	256	323	310	254	193	146	113	81	61	46	36	31	27	24	22	20	18	16	12	1
.40	8	11	14	19	21	23	27	32	42	61	91	132	181	276	318	294	237	181	138	95	70	51	39	32	28	25	23	21	18	16	12	1
.50	8	10	13	18	20	22	25	30	38	53	78	114	159	253	311	300	251	195	149	102	74	53	40	33	29	25	23	21	16	15	12	1
.75	7	8	11	14	16	19	21	25	30	38	53	76	146	256	284	256	208	143	99	66	46	36	31	27	24	22	20	19	17	16	12	2
1.0	5	7	9	12	12	13	14	17	19	22	25	31	57	111	188	256	286	272	208	144	90	56	46	41	33	29	26	23	20	17	13	4
1.5	4	5	6	8	8	9	10	11	12	13	14	15	17	22	33	59	107	171	231	268	235	157	88	56	41	33	29	25	21	18	14	7
2.0	2	3	4	5	5	6	6	7	7	8	9	9	10	12	15	19	27	44	78	157	252	167	96	59	42	34	29	25	23	20	15	11
2.5	1	2	2	4	4	4	5	5	6	6	7	7	8	10	12	15	19	27	58	120	214	241	159	94	59	42	34	29	21	16	12	4
3.0	0	1	1	5	2	3	4	4	4	4	5	5	6	7	8	10	12	14	22	44	113	214	214	152	91	58	42	33	23	17	12	2

IA/P = 0.30

TRVL TIME (HR)	11.0	11.3	11.6	11.9	12.0	12.1	12.2	12.3	12.4	12.5	12.6	12.7	12.8	13.0	13.2	13.4	13.6	13.8	14.0	14.3	14.6	15.0	15.5	16.0	16.5	17.0	17.5	18.0	19.0	20.0	22.0	26.0
0.0	0	0	1	1	1	4	16	42	83	137	195	243	271	292	227	178	143	117	98	79	66	55	47	42	38	34	31	30	27	23	19	13
.10	0	0	0	0	0	0	3	12	32	66	113	168	218	279	260	213	169	136	113	88	72	59	49	43	39	35	32	30	27	23	19	17
.20	0	0	0	0	0	2	6	9	24	52	93	143	193	271	256	225	180	145	119	92	75	60	50	44	39	35	32	30	24	24	19	18
.30	0	0	0	0	0	1	3	6	18	41	75	120	169	246	264	234	191	153	125	96	78	62	51	44	40	36	33	31	27	19	19	18
.40	0	0	0	0	0	1	4	14	32	61	100	146	190	259	222	181	146	109	86	67	53	46	41	37	34	31	28	25	22	—	—	—
.50	0	0	0	0	0	0	0	10	24	49	83	168	191	230	254	237	191	155	115	90	69	55	47	42	37	34	29	28	25	—	—	—
.75	0	0	0	0	0	0	0	0	0	2	20	25	76	213	239	228	198	149	112	82	61	50	44	40	35	35	34	29	20	—	—	—
1.0	0	0	0	0	0	0	0	0	0	0	0	2	15	113	182	226	234	197	150	104	72	56	47	42	38	34	30	27	20	—	—	—
1.5	0	0	0	0	0	0	0	0	0	0	0	1	4	18	51	104	162	220	221	187	158	102	71	56	47	42	37	28	22	—	—	13
2.0	0	0	0	0	0	0	0	0	0	0	0	0	0	1	5	20	49	121	209	152	100	70	55	47	41	34	29	23	—	—	17
2.5	0	0	0	0	0	0	0	0	0	0	0	0	0	0	2	7	32	87	171	199	146	98	69	54	46	37	35	31	24	—	18
3.0	0	0	0	0	0	0	0	0	0	0	0	0	0	0	0	0	2	13	62	158	192	151	103	73	58	42	41	34	26	—	18

IA/P = 0.50

TRVL TIME (HR)	11.0	11.3	11.6	11.9	12.0	12.1	12.2	12.3	12.4	12.5	12.6	12.7	12.8	13.0	13.2	13.4	13.6	13.8	14.0	14.3	14.6	15.0	15.5	16.0	16.5	17.0	17.5	18.0	19.0	20.0	22.0	26.0
0.0	7	11	16	31	71	101	126	123	154	160	138	110	100	87	77	67	60	55	50	46	43	41	45	54	59	46	43	41	38	34	28	17
.10	0	0	0	0	33	58	87	134	156	149	120	108	93	82	71	62	57	52	50	47	44	42	46	55	65	50	46	42	39	36	29	24
.20	0	0	0	0	26	48	74	137	153	140	123	111	95	84	72	63	57	52	47	44	42	40	44	52	62	47	44	42	39	36	29	27
.30	0	0	0	0	20	38	62	111	150	143	127	114	98	86	73	63	58	53	48	45	41	42	46	53	59	48	45	42	35	35	29	27
.40	0	0	0	0	6	16	31	75	148	145	137	123	106	91	77	66	59	54	49	45	43	43	45	54	59	49	45	43	39	35	29	29
.50	0	0	0	0	2	12	25	64	137	139	136	127	108	94	79	67	60	55	50	46	43	43	46	55	60	50	46	43	39	36	29	29
.75	0	0	0	0	0	2	5	40	115	136	140	134	117	101	84	70	62	56	51	47	44	40	47	51	78	47	44	41	37	36	36	—
1.0	0	0	0	0	0	0	1	3	59	96	125	139	133	117	97	78	66	59	54	49	46	66	78	59	54	54	49	46	41	37	—	8
1.5	0	0	0	0	0	0	0	0	0	0	9	26	54	86	123	133	119	95	77	66	59	54	77	95	66	59	54	49	43	39	31	17
2.0	0	0	0	0	0	0	0	0	0	0	0	3	10	54	108	129	116	93	76	67	60	55	76	93	65	60	58	53	45	41	33	24
2.5	0	0	0	0	0	0	0	0	0	0	0	0	2	10	34	84	125	117	96	78	66	62	66	96	78	62	57	49	43	37	35	27
3.0	0	0	0	0	0	0	0	0	0	0	0	0	0	1	6	32	89	114	114	97	78	66	59	94	77	54	48	46	41	37	—	27

Figure D-6. (Chart 3)   Tabular hydrograph unit discharges (csm/in) for type II rainfall distribution (t_c = 1.0 hour). *(Courtesy of Soil Conservation Service, Technical Release 55.)*

HYDROGRAPH TIME (HOURS)

IA/P = 0.10   * * * TC = 1.25 HR * * *

TRVL TIME (HR)	11.0	11.3	11.6	11.9	12.0	12.1	12.2	12.3	12.4	12.5	12.6	12.7	12.8	13.0	13.2	13.4	13.6	13.8	14.0	14.3	14.6	15.0	15.5	16.0	16.5	17.0	17.5	18.0	19.0	20.0	22.0	26.0
0.0	10	13	18	25	29	38	54	81	118	163	213	256	284	311	266	212	163	129	104	78	61	47	37	31	27	24	22	20	18	16	12	1
.10	10	13	17	23	27	34	47	69	102	143	189	234	267	297	274	226	175	138	111	82	64	48	38	31	27	24	22	20	18	16	12	1
.20	9	11	15	20	22	26	31	42	60	88	124	168	212	261	292	280	212	166	131	95	72	53	40	33	28	25	23	21	18	16	12	2
.30	8	11	14	19	21	24	29	38	53	76	108	148	190	263	288	268	224	177	140	101	76	55	41	34	30	26	23	21	18	16	12	2
.40	8	10	13	18	20	23	27	34	46	66	94	130	170	245	282	273	235	188	149	107	80	58	42	34	29	26	23	21	19	16	12	2
.50	7	9	12	16	17	19	22	25	31	41	58	82	114	190	256	279	262	222	178	127	93	65	46	36	31	27	24	23	19	17	13	3
.75	6	8	10	14	15	17	19	22	25	31	41	56	78	139	207	254	265	245	208	152	110	75	51	39	32	28	25	22	20	18	14	3
1.0	5	6	8	10	11	13	14	15	17	19	22	26	33	60	109	173	230	261	255	208	153	100	64	46	36	30	26	24	20	18	16	3
1.5	3	4	5	7	7	8	9	10	11	12	15	19	27	45	79	130	186	239	247	180	108	68	48	37	31	27	24	22	19	14	10	10
2.0	2	3	4	5	6	7	7	8	8	8	9	10	13	22	35	59	98	171	236	156	95	62	44	35	30	23	20	17	15	16	11	11
2.5	1	1	2	3	3	4	4	5	5	5	6	6	8	14	19	28	58	114	197	226	163	102	65	46	36	28	21	17	16	18	11	11
3.0	0	1	1	1	2	2	3	3	3	3	4	4	5	10	13	19	35	88	184	208	169	109	70	55	46	30	24	20	18	12	12	12

IA/P = 0.30   * * * TC = 1.25 HR * * *

TRVL TIME (HR)	11.0	11.3	11.6	11.9	12.0	12.1	12.2	12.3	12.4	12.5	12.6	12.7	12.8	13.0	13.2	13.4	13.6	13.8	14.0	14.3	14.6	15.0	15.5	16.0	16.5	17.0	17.5	18.0	19.0	20.0	22.0	26.0
0.0	0	0	2	9	25	50	86	130	174	208	253	235	201	164	136	115	92	76	61	51	44	39	35	32	30	27	24	19	17	2
.10	0	0	0	1	7	19	40	71	110	153	235	247	227	191	157	136	108	84	66	53	46	41	36	33	31	28	24	19	19	2
.20	0	0	0	1	4	14	31	58	93	202	231	199	165	138	108	87	68	55	47	41	37	34	31	28	25	21	20	3
.30	0	0	0	3	10	24	46	77	152	236	222	190	158	122	97	74	58	49	43	38	34	32	28	25	21	20	3
.40	0	0	0	2	8	19	37	64	134	196	232	198	166	127	101	77	59	50	43	38	35	32	28	25	22	25	3
.50	0	0	0	0	2	9	14	30	82	151	206	217	189	146	113	85	64	52	45	40	36	33	29	26	23	6
.75	0	0	0	0	1	2	7	15	49	105	164	218	205	166	129	95	69	55	47	41	37	32	29	27	10
1.0	0	0	0	0	0	0	0	1	9	32	77	134	185	203	166	120	83	63	52	45	39	35	32	15
1.5	0	0	0	0	0	0	0	0	2	11	33	72	121	184	117	82	62	51	44	39	35	29	22	18
2.0	0	0	0	0	0	0	0	0	0	2	7	21	67	132	174	123	86	64	52	45	31	24	18
2.5	0	0	0	0	0	0	0	0	0	1	3	13	46	121	187	166	119	84	63	52	39	25	18
3.0	0	0	0	0	0	0	0	0	0	0	1	8	44	129	180	160	116	83	63	44	27	18

IA/P = 0.50   * * * TC = 1.25 HR * * *

TRVL TIME (HR)	11.0	11.3	11.6	11.9	12.0	12.1	12.2	12.3	12.4	12.5	12.6	12.7	12.8	13.0	13.2	13.4	13.6	13.8	14.0	14.3	14.6	15.0	15.5	16.0	16.5	17.0	17.5	18.0	19.0	20.0	22.0	26.0
0.0	0	0	0	1	5	13	26	44	68	91	125	142	128	117	107	94	83	72	63	56	52	47	44	42	38	34	28	36	40	52	5
.10	0	0	0	0	0	7	16	20	36	57	100	142	136	114	88	76	65	57	54	49	45	43	39	35	29	36	42	56	6
.20	0	0	0	0	0	3	2	15	24	59	90	126	117	102	90	78	67	60	56	54	49	45	41	35	30	37	44	62	9
.30	0	0	0	0	0	1	3	9	24	98	139	137	125	109	96	82	71	62	56	51	46	44	40	36	29	38	47	70	13
.40	0	0	0	0	0	1	1	3	17	40	71	101	134	127	112	98	83	72	63	57	51	47	44	40	36	44	52	—	—
.50	0	0	0	0	0	0	0	3	15	43	79	112	135	129	103	85	71	63	56	52	47	44	40	37	44	56	—	—
.75	0	0	0	0	0	0	0	1	3	15	39	71	123	130	112	94	78	67	60	54	46	41	38	62	—	—
1.0	0	0	0	0	0	0	0	0	0	4	17	40	101	129	121	103	84	71	62	56	51	47	42	70	—	—
1.5	0	0	0	0	0	0	0	0	0	0	3	10	26	92	119	125	105	86	72	63	57	52	44	—	—
2.0	0	0	0	0	0	0	0	0	0	0	0	3	11	35	72	112	122	103	85	71	63	56	47	—	—
2.5	0	0	0	0	0	0	0	0	0	0	0	0	3	24	66	111	119	101	83	71	62	51	—	—
3.0	0	0	0	0	0	0	0	0	0	0	0	0	0	4	23	71	110	116	101	82	70	55	—	—

RAINFALL TYPE = II

Figure D-6. (Chart 4) Tabular hydrograph unit discharges (csm/in) for type II rainfall distribution (t_c = 1.25 hours). *(Courtesy of Soil Conservation Service, Technical Release 55.)*

--- HYDROGRAPH TIME (HOURS) ---

RAINFALL TYPE = II   * * * TC = 1.5 HR * * *

IA/P = 0.10

TRVL TIME (HR)	11.0	11.3	11.6	11.9	12.0	12.1	12.2	12.3	12.4	12.5	12.6	12.7	12.8	13.0	13.2	13.4	13.6	13.8	14.0	14.3	14.6	15.0	15.5	16.0	16.5	17.0	17.5	18.0	19.0	20.0	22.0	26.0
0.0	9	11	15	21	25	31	41	58	82	112	147	184	216	255	275	236	198	159	129	98	76	57	43	35	30	25	23	21	18	16	12	1
.10	8	10	13	18	20	23	28	37	51	72	98	131	166	226	265	254	226	187	151	113	86	63	46	37	31	25	23	21	16	13	12	2
.20	8	10	13	17	19	22	26	33	45	63	87	116	149	212	259	259	233	197	160	119	90	66	48	38	32	27	24	22	16	13	13	2
.30	7	9	12	16	18	21	24	30	40	55	76	103	134	197	244	255	238	206	169	125	95	68	49	38	32	27	24	22	17	13	13	2
.40	7	8	11	14	15	17	19	23	28	36	49	67	91	151	208	247	252	230	196	146	109	77	54	41	34	29	25	22	17	13	13	3
.50	6	7	10	13	15	16	18	21	26	33	43	59	80	136	194	238	249	235	204	154	115	81	56	42	34	29	25	23	17	14	13	3
.75	5	7	8	11	13	16	18	21	25	32	42	57	67	101	172	218	233	240	233	193	148	102	67	48	38	34	30	26	18	16	14	5
1.0	4	5	7	9	11	12	13	16	18	21	25	32	42	76	125	148	201	230	236	204	193	135	86	59	44	35	31	30	18	17	14	7
1.5	3	4	5	6	6	7	8	8	9	10	11	12	13	16	22	34	58	95	141	203	226	197	131	84	58	43	35	29	23	20	15	10
2.0	1	2	3	4	4	5	5	6	6	7	7	9	10	12	16	22	34	56	110	172	218	187	126	82	57	43	34	29	25	21	16	11
2.5	1	1	2	2	2	3	3	3	4	4	5	5	6	7	8	9	14	18	34	69	141	210	190	133	87	60	44	35	23	17	12
3.0	0	0	1	2	2	2	3	3	3	4	5	6	8	9	11	16	27	66	149	204	181	128	85	58	44	37	35	28	19	16	12

IA/P = 0.30

TRVL TIME (HR)	11.0	11.3	11.6	11.9	12.0	12.1	12.2	12.3	12.4	12.5	12.6	12.7	12.8	13.0	13.2	13.4	13.6	13.8	14.0	14.3	14.6	15.0	15.5	16.0	16.5	17.0	17.5	18.0	19.0	20.0	22.0	26.0
0.0	0	0	0	1	1	6	15	31	53	80	112	144	193	225	208	186	157	134	108	89	70	56	48	42	37	34	31	28	25	20	20	2
.10	0	0	0	0	1	4	12	25	43	68	97	157	208	219	203	178	151	131	117	99	77	60	49	44	38	35	32	29	25	20	20	3
.20	0	0	0	0	0	3	9	19	35	57	114	168	201	213	196	171	135	108	84	64	53	46	40	36	33	29	26	21	20	20	20	4
.30	0	0	0	0	1	2	7	15	29	48	100	155	193	210	200	177	140	113	87	66	54	48	46	41	36	33	29	26	20	20	20	5
.40	0	0	0	0	2	5	12	23	39	87	117	182	207	202	184	146	117	89	73	68	66	64	62	56	48	41	36	33	26	20	20	5
.50	0	0	0	0	0	6	13	26	51	97	131	164	190	205	178	153	99	77	73	73	72	64	60	58	49	43	35	32	28	25	21	7
.75	0	0	0	1	4	16	30	57	80	110	160	189	197	187	147	110	80	80	89	97	97	74	62	56	45	38	34	31	29	25	21	8
1.0	1	7	21	47	85	97	99	45	13	5	1	0	138	175	178	137	97	57	31	42	72	133	129	95	72	48	37	31	28	25	21	12
1.5	0	0	0	0	0	0	0	1	13	45	85	145	187	178	97	47	13	4	0	0	0	95	161	174	133	97	72	58	38	32	25	16
2.0	0	0	0	0	0	0	0	0	0	0	0	0	0	1	7	21	47	85	97	45	31	29	89	126	174	133	98	72	48	34	26	18
2.5	0	0	0	0	0	0	0	0	0	0	0	0	0	0	0	0	1	11	32	63	48	15	4	3	9	87	169	111	37	27	23	18
3.0	0	0	0	0	0	0	0	0	0	0	0	0	0	0	0	0	0	9	4	0	0	0	0	0	128	129	95	54	34	26	19

IA/P = 0.50

TRVL TIME (HR)	11.0	11.3	11.6	11.9	12.0	12.1	12.2	12.3	12.4	12.5	12.6	12.7	12.8	13.0	13.2	13.4	13.6	13.8	14.0	14.3	14.6	15.0	15.5	16.0	16.5	17.0	17.5	18.0	19.0	20.0	22.0	26.0
0.0	0	0	0	0	0	0	8	16	27	42	59	92	116	128	130	128	121	112	102	90	78	67	60	55	50	46	43	43	39	35	29	4
.10	0	0	0	0	0	0	6	12	22	35	51	84	110	121	128	130	123	114	102	91	79	68	61	55	50	46	43	43	39	35	29	4
.20	0	0	0	0	0	0	4	10	18	29	44	71	91	114	128	128	126	120	108	97	83	71	63	57	52	47	44	44	40	36	29	5
.30	0	0	0	0	0	0	3	8	14	24	36	63	83	108	126	126	122	122	110	98	85	72	63	57	52	48	44	44	40	36	29	6
.40	0	0	0	0	0	2	6	12	20	31	49	75	90	112	124	126	116	116	116	104	90	78	66	59	54	49	45	45	41	37	29	8
.50	0	0	0	0	0	2	5	12	20	35	53	77	91	106	121	125	128	123	112	107	91	79	67	60	54	49	46	46	41	37	29	8
.75	0	0	0	0	1	5	12	22	36	57	81	104	108	114	120	128	122	120	112	108	97	81	71	63	56	51	47	47	42	38	30	11
1.0	0	0	1	1	0	0	2	2	0	0	24	52	83	108	110	108	122	118	122	110	108	90	72	66	59	54	49	49	43	39	31	16
1.5	0	0	0	0	0	0	0	0	0	1	3	6	11	25	45	80	112	115	118	112	90	75	59	65	54	49	45	41	45	41	32	23
2.0	0	0	0	0	0	0	0	0	0	0	0	1	1	4	11	32	63	104	100	83	91	77	60	54	47	48	46	42	42	42	34	26
2.5	0	0	0	0	0	0	0	0	0	0	0	0	0	1	4	16	45	97	104	88	69	81	113	103	94	56	51	53	45	38	30	27
3.0	0	0	0	0	0	0	0	0	0	0	0	0	0	0	0	3	10	26	49	75	98	108	111	90	54	59	54	48	43	36	28	28

Figure D-6. (Chart 5)   Tabular hydrograph unit discharges (csm/in) for type II rainfall distribution (t_c = 1.5 hours). *(Courtesy of Soil Conservation Service, Technical Release 55.)*

Figure D-6. (Chart 6) Tabular hydrograph unit discharges (csm/in) for type II rainfall distribution (t_c = 2.0 hours). *(Courtesy of Soil Conservation Service, Technical Release 55.)*

E

COMPUTER SOFTWARE APPLICATIONS FOR STORMWATER MANAGEMENT (SELECTED LIST)

The following list consists of a selected sample of available software for use in stormwater management applications taken from various sources. A major source is the special Buyers' Guide for Hydrology/Hydraulics Software published in *Civil Engineering News*. The list is arranged by sub-topic.

SOFTWARE NAME	AVAILABLE FROM
Watershed Modeling	
HEC-1	Corps of Engineers Davis, CA (Also available through many vendors)
HydroCAD	Applied Microcomputer Systems Chocorua, NH
Watershed Modeling	Eagle Point Dubuque, IA
Hydraflow Hydrographs for Windows	Intellsolve Kennesaw, GA
AutoCivil	Research Engineers Yorba Linda, CA
WIMP	Simplicity Systems, Inc. East Grand Forks, MN
Storm Sewer System	
STORM	Corps of Engineers Davis, CA
HydroWorks PM	Applied Geographic Technologies, Inc. Ft. Worth, TX

SewerWORKS	BOSS International Madison, WI
SewerDESIGN	BOSS International Madison, WI
Storm Sewers	Eagle Point Dubuque, IA
CIA	Engenious Systems, Inc. Seattle, WA
StormCAD	Haestad Methods, Inc. Waterbury, CT
Hydraflow Storm	Intellsolve Kennesaw, GA
HydraGraphics	Pizer, Inc. Seattle, WA
TERRAMODEL Sewer Design & Analysis Module	Plus 3 Software, Inc. Atlanta, GA
HydroPak	Simplicity Systems, Inc. East Grand Forks, MN
NETWORX	Streamline Technologies, Inc. Winter Park, FL
XP-RAT HGL	XP Software Tampa, FL

Culvert Design

WSPRO	BOSS International Madison, WI
HYDROCALC (Hydraulics)	Dodson & Assoc., Inc. Houston, TX
Culvert Master for Windows	Haestad Methods, Inc. Waterbury, CT

Detention Basin Design

TR20	Natural Resoures Conservation Service Washington, DC
HYDROCALC (Hydrology)	Dodson & Assoc., Inc. Houston, TX
Water Works	Engenious Systems, Inc. Seattle, WA

Stormwater Management

SWMM	EPA Washington, DC
StormMaster	PacSoft, Inc. Bothell, WA
TERRAMODEL Hydrology Module	Plus 3 Software, Inc. Atlanta, GA
ICPR—Interconnected Channel and Pond Routing Model	Streamline Technologies, Inc. Winter Park, FL
ICAD	Streamline Technologies, Inc. Winter Park, FL
XP-SWMM	XP Software Tampa, FL

Water Surface Profile (Backwater)

HEC-2	Corps of Engineers Davis, CA
HEC-RAS	BOSS International Madison, WI
	West Consultants, Inc. Carlsbad, CA
	Dodson & Assoc., Inc. Houston, TX
Water Surface Profiling	Eagle Point Dubuque, IA
Hydraflow Open Channels	Intellsolve Kennesaw, GA

F

GLOSSARY

The following terms are defined as they are used in this text.

Absolute Viscosity—A measure of the influence of the motion of one layer of a fluid upon another layer a short distance away.

Accuracy—A calculated value close to the actual value of a quantity is said to be accurate. In good engineering practice, accuracy of measurement is more important than precision.

Adhesion—A property of water which allows it to cling to another body.

Apron—Concrete or riprap lining of the ground at the inlet or outlet of a storm sewer or culvert.

Attenuation—A reduction of rate of flow accomplished by a detention basin, which temporarily stores stormwater and then releases it slowly. Attenuation can also refer to the alteration in a hydrograph which occurs as water flows downstream.

Backwater Curve—A variation in the water surface profile caused by an obstruction, such as a bridge crossing.

Base Flow—The constant low-level flow in streams due to subsurface feed.

Basin Divide—*See* Divide

Berm—Earth embankment constructed on the downhill side of a detention basin to help contain the stored water.

Bernoulli Equation—Equation formulating the conservation of energy in hydraulics.

Broad-crested Weir—A commonly employed weir for dams and detention basins. The multi-stage weir, a variation of the broad-crested weir, is used to very precisely regulate discharge.

Buoyancy—The uplifting force exerted by water on a submerged solid object.

Calculation—The overall presentation of a mathematical solution, including computations and explanations of the computations.

Catch Basin—Storm water inlet structure with sediment trap at bottom.

Catchment Area—Synonym for drainage basin or watershed.

Cipoletti Weir—A trapezoidal variation of the sharp-crested weir devised to compensate for loss of flow quantity due to contractions at the vertical edges of a rectangular weir.

Civil Engineer—An engineer specializing in the design and construction of structures and public works.

Cleansing Velocity—The minimum velocity of water flow through a pipe to avoid deposits of silt and debris.

CMP—Corrugated metal pipe, used for storm sewers and culverts composed of aluminum or steel.

Cohesion—A property of water which allows it to resist a slight tensile stress.

Combination Drain—A storm sewer made of perforated pipe backfilled with gravel to intercept ground water in addition to conveying stormwater.

Combined Sewer—An obsolete system once used to convey both sewage waste and storm water through the same pipes. Combined sewers have been nearly eliminated in favor of separate storm and sanitary sewers.

Computation—For the purposes of this text, computation refers to the mathematical manipulation of numbers.

Confluence—The intersection of two branches of a stream.

Continuity Equation—There are two different continuity equations in use in hydraulics. The first is a measure of the relationship between the quantity of water, its rate of flow, and the cross-sectional area it is flowing through, expressed most simply as $Q = va$. The second is used in routing, and is a statement of the conservation of mass of the water entering and leaving a reservoir or impoundment.

Contraction—A loss of energy which takes place as water flows past the vertical sides of a weir.

CPP—Corrugated plastic pipe, used for storm sewers and culverts composed of high density polyethylene.

Critical Depth—Depth of flow in an open channel for which specific energy is at minimum value.

Critical Slope—A slope that causes normal depth to coincide with critical depth.

Critical Velocity—The velocity of water at critical depth.

Crown—The top of the inside of a pipe or culvert.

Culvert—Conduit to convey a stream or runoff through an embankment.

Density—A measure of mass per unit of volume.

Design—The process of determining the complete specification of a structure. Design is a complex and ever-growing intellectual endeavor, which requires the incorporation of numerous criteria following good engineering practice.

Design Storm—The design storm is the largest storm expected to occur in a given period of time specified in the design parameters for a hydraulic project.

Detention Basin—A facility designed to temporarily store stormwater during a rainfall event and then release the water at a slow rate.

Discharge—Flow of stormwater either overland or in a conduit; measured as a rate of flow in cfs (m^3/s).

Discharge Coefficient—Used in computing the rate of discharge over a rectangular weir, the discharge coefficient is a dimensionless constant which accounts for a number of various hydraulic factors.

Discharge Rating—A table important in detention basin design, which displays discharge through the outlet structure as a function of water level in the impoundment.

Divide—Also called basin divide or watershed divide, it is the line on a map that outlines the watershed. All rainfall landing outside the divide does not flow to the point of analysis.

Drainage Basin—Area of land over which rainfall flows by gravity to a single point called the point of analysis or point of concentration.

Emergency Spillway—A safety feature to prevent detention basins from overflowing, consisting of an additional outlet set higher than the other outlets. Water will only enter if the impoundment level rises higher than anticipated in the original design.

Energy Dissipater—A specially designed obstruction block placed at an outlet to create head loss in very high velocity situations.

Energy Grade Line—Line graphed along a one-dimensional hydraulic system representing total energy or total head at every point along the system.

Entrance Coefficient—Used in computing orifice flow, the entrance coefficient is a dimensionless proportionality constant which accounts for the reduction of flow due to entrance head loss.

Entrance Loss—A small but sudden drop of the energy grade line at the point where water enters a pipe from a larger body of water caused by loss of energy as the turbulent water enters the more restrictive pipe.

Erosion Control Mat—Mesh netting placed on the ground which will protect the ground surface and anchor a vegetative cover.

Evapotranspiration—When rainfall strikes the ground, some of it is absorbed by plants and some evaporates immediately. This lost rainwater is not available for runoff.

First Flush—The first flow of runoff which picks up the loose dust that has coated the ground since the last rainfall. First flush is responsible for most transport of pollutants.

Flared End Section—A precast section of pipe made in a flared shape to use at the inlet or outfall end of a storm sewer or culvert in place of a headwall.

Flood Flow—A quick surge in stream flow due to runoff from a rainfall event.

Freeboard—The vertical distance from the maximum or design level of water to the top of the structure containing or conveying the water.

Free Surface—The surface of water in a container or conduit which is exposed to the atmosphere.

Friction Loss—The constant drop in energy of water as it flows along the length of a pipe caused by contact with the inside surface of the pipe.

Gabion—A rectangular wire mesh basket filled with rocks and placed on the ground as a protective lining similar to riprap.

Gauge Pressure—Water pressure in excess of atmospheric pressure.

Gradation—The use of stones in a variety of sizes which promotes interlocking. Gradation is important to the overall stability of riprap design.

Gradient—A measure of the slope of a channel or conduit expressed in ft/ft (m/m).

Groundwater—A large pool of underground water filling the voids between soil and rock particles.

Head—Energy per unit weight of water expressed as feet (m). It may be potential or kinetic energy or it may be a loss of energy in some other form.

Head Loss—Loss of energy (expressed as energy per unit weight of water) due to friction or obstruction such as entrance to a culvert.

Headwall—Short retaining wall placed at the end of a storm sewer pipe (inlet or outlet).

Headwater Depth—Upstream water depth, which provides the potential energy to drive water through a culvert. It can therefore become a measure of the capacity of a given culvert.

Hydraulic Grade Line—A line drawn along a one-dimensional hydraulic system depicting potential energy expressed as position plus pressure head at all points along the system.

Hydraulic Path—Path traveled by the drop of rain water landing at the most remote point in the drainage basin as it flows to the point of analysis.

Hydraulic Radius—Mathematical term defined by the ratio of cross-sectional area to wetted perimeter; it is a measure of the hydraulic efficiency of an open channel.

Hydraulics—The study of the mechanics of water and other fluids at rest and in motion.

Hydrograph—A graph of runoff quantity or discharge versus time at the point of analysis of a drainage basin.

Hydrologic Cycle—The natural pattern of water evaporation, condensation, precipitation and flow.

Hydrology—The study of rainfall and the subsequent movement of rainwater including runoff.

Hydrostatics—The study of water at rest.

Infiltration Basin—A detention basin which promotes recharge of stormwater to groundwater storage.

Impervious—A ground cover condition in which no rainfall infiltrates into the ground. Typically, pavement and roofs are considered impervious.

Impoundment—Volume of water stored in a detention basin.

Infiltration Rate—The rate at which rainwater is absorbed by the ground.

Inflow—Rate of flow of stormwater into a detention basin.

Initial Losses—Also called initial abstraction, it is the amount of rainfall in inches that infiltrates into the ground before any runoff begins.

Inlet—A component of a storm sewer system, it is a precast concrete structure placed in the ground with a cast iron grate on the top. Stormwater enters through the grate and into the storm sewer system.

Intensity-Duration-Frequency (IDF) Curve—Central to the Rational Method for determining peak runoff, IDF curves are developed by various government agencies based upon data from Weather Bureau records. They show the relationship between rainfall intensity and duration for various return periods at given locations in the U.S.

Invert—Lowest point on the cross section of a conduit, such as a pipe, channel or culvert.

Irrigation—Supplying water for crops by artificial means, such as conveying it through channels and pipes.

Iteration—A mathematical process used to solve an equation indirectly by trial and error.

Kinematic Viscosity—Absolute viscosity divided by density; a measurement useful in hydraulic problems affected by density.

Lag—A parameter used in runoff analysis; the time separation between the centroid of the rainfall excess graph and the peak of the hydrograph. Lag is related empirically to time of concentration.

Laminar Flow—Smooth, non-turbulent flow of water in a conduit, usually having low velocity.

Land Development—Constructing man-made structures on a parcel of land for use by people. It usually results in an increase in impervious surfaces and therefore an increase in runoff.

Metric System—Also known as the International System of Units (SI), the metric system is used as a standard of measurement by most of the world. The United States has not yet completed a changeover to metric units, so the civil engineer is faced with the need to be conversant with both metric and English systems of measurement.

Normal Depth—Vertical distance from the invert of a channel or conduit to the free water surface when water is flowing without the influence of backwater.

Ogee Weir—A rectangular weir used commonly as a spillway; the smooth rounded surface of the ogee weir is designed to reduce energy loss by contraction.

Open Channel Flow—Water that flows downhill through a conduit with its surface exposed to the atmosphere is said to undergo open channel flow.

Orifice—An opening in a container through which stored water may flow.

Overbank—Land immediately adjacent to each side of the channel of a stream. When flow in the stream exceeds the top of bank, it spills onto the overbank area.

Overland Flow—Stormwater runoff flowing over the ground surface in the form of sheet flow. It usually occurs at the beginning of the hydraulic path.

Outfall—The point of a storm sewer system where discharge leaves the system to enter the receiving body of water.

Outflow—Rate of flow of stored water out of a detention basin.

Outlet—The downstream end of a culvert.

Piezometer—A simple device for measuring water pressure, both static and dynamic.

Pipe—A hollow cylinder used in storm sewer systems to convey stormwater toward a receiving stream.

Planimeter—A device for measuring the area contained within a line making a closed figure on a map.

Point of Analysis—Any point on the ground or along a stream at which the quantity of runoff from the upstream catchment area is to be determined.

Point of Concentration—Synonym for point of analysis.

Precipitation—Water which lands on the ground from the sky as part of the hydrologic cycle.

Precision—Fineness of measurement expressed by the number of significant figures. Precise measurements are useless if they are not accurate.

Pressure—Force exerted by water against a unit area caused by the weight of the water above the point.

PVC—Polyvinyl chloride pipe usually used for roof drains and sanitary sewer mains.

Rainfall Excess—The remainder of rainfall that reaches the point of analysis after initial losses and infiltration.

RCP—Reinforced concrete pipe, used for storm sewers and culverts.

Receiving Water—A body of water into which a storm sewer system discharges. It usually consists of a stream, lake or another storm sewer.

Recharge—Engineered effort to re-direct some runoff into the ground where it would seep down to the groundwater.

Retention Basin—Term used for a detention basin which holds some water under normal non-flooding conditions. It serves as a sediment basin as well as a detention basin.

Return Period—The average number of years between two rainfall events that equal or exceed a given number of inches over a given duration.

Ridge—A land formation shaped so that runoff diverges as it flows downhill. It is the opposite of a swale. A drainage basin divide runs along a ridge.

Riprap—Stones placed on the ground used as a protective lining at inlets and outlets of storm sewers and culverts.

Roughness Factor—A dimensionless parameter used in Manning's Equation describing the roughness of the surface of a channel or pipe.

Routing—In hydraulics, a mathematical procedure for computing an outflow hydrograph when the inflow hydrograph is known.

Runoff—The quantity of surface flow resulting from rainfall.

Runoff Coefficient—A dimensionless proportionality factor used in the Rational Method to account for infiltration and evapotranspiration.

Saddle—A topographic feature marking the transition between two ridges and two swales.

Sediment Basin—A depression in the ground lower than the invert of its outlet so that stormwater passing through will deposit its silt and sediment in the resulting ponded water.

Shallow Concentrated Flow—The form taken by stormwater runoff as it flows along the ground and converges into rivulets due to irregularities in the ground surface.

Sheet Flow—The form taken by stormwater as it flows along a smooth flat surface. Flow lines remain parallel and do not converge.

Significant Figures—The number of significant figures in a quantity is the number of digits minus the number of zeros used only as place holders. The number of significant figures in the least precisely measured quantity involved in a calculation determines the precision of the result.

Slope—A quantity indicating the extent to which a channel or conduit tips along its long axis. The slope is the ratio of the vertical drop to the length multiplied by 100 and expressed as a percent.

Specific Energy—A mathematical formulation equal to the flow depth in an open channel plus the velocity head. It is the total energy head above the channel bed.

Specific Weight—The weight of water per unit volume. The specific weight of water is taken as 62.4 lb/ft³ (9.8 x 10³ N/m³).

Spillway—A structure for regulating the outflow from a reservoir or detention basin. It generally consists of a weir or orifice or both.

Spring Line—A line running the length of a pipe midway up the cross section of the pipe.

Stage—Term used as a synonym for elevation of water level in a stream or detention basin.

Steady Flow—When the rate of flow does not significantly vary with respect to time at any point along a stream, its flow is said to be steady.

Stilling Basin—A depression in the ground surface at an outlet designed to trap water and to absorb excessive energy of discharge.

Storm Sewer—A pipe, usually underground, used to convey stormwater runoff.

Stormwater—Water which falls to earth as precipitation and then runs along the ground impelled by gravity.

Stormwater Management—Term used to describe all endeavors to control the quantity and quality of runoff in areas affected by land development.

Stream Rating Curve—A graph of discharge versus water surface elevation used to analyze stream flow.

Sub-basin—A portion of a drainage basin functioning as a complete drainage basin when calculating runoff. If a drainage basin is not homogenous, it should be partitioned into two or more sub-basins.

Subcritical—When flow depth is greater than critical depth, the flow is relatively tranquil and is called subcritical.

Subsurface Flow—Water which originates as precipitation and infiltrates a short distance into the ground, then runs laterally within the ground eventually reaching a stream.

Supercritical—Flow depths below critical depth flow rapidly and are called supercritical.

Superposition—A principle used to add two or more hydrographs to obtain the resulting total hydrograph.

Swale—A land formation shaped so that runoff converges as it flows downhill. It is the opposite of a ridge. A basin divide never runs along a swale.

Tailwater—water leaving a storm sewer or culvert.

Time of Concentration—The amount of time needed for runoff to flow from the most remote point in the drainage basin to the point of analysis.

Trash Racks—A grill composed of steel or some other suitable material placed over the outlet or spillway of a detention basin to prevent clogging by trash or debris.

Turbulent Flow—Water flowing at a great enough velocity to develop eddies is said to be turbulent. Turbulent flow results in more friction loss.

Unit Hydrograph—A generalized hydrograph resulting from a rainfall excess of one unit (1 inch or 1 cm).

Unsteady Flow—When there is a change in the rate of flow at any point along a stream, the water's flow is said to be unsteady.

Viscosity—The ability of fluid molecules to flow past each other.

Water Hammer—The extreme variation in pressure within a pipe caused by an abrupt stoppage in flow.

Watershed—Synonym for drainage basin.

Weir—A structure, usually horizontal, placed in a stream or pond over which water flows.

Wet Basin—A detention basin constructed to have a permanent pond at its bottom. Also called a retention basin or sediment basin, it is especially effective at trapping pollutants.

Wetlands—Environmentally important and vulnerable land characterized by high groundwater and/or standing water and wet-tolerant vegetation. Land development is severely restricted on and near such land.

Wetted Perimeter—The distance along the cross-section of a channel or conduit where the surface is in contact with flowing water.

Wingwall—Retaining wall placed at each end of a culvert to stabilize the embankment slope and help drect the flow.

APPENDIX

G

SYMBOLS

The following symbols representing mathematical quantities are defined as they are used in this text. Units are given in the English system followed by metric units in parentheses. Parameters used in the SCS Method are defined in English units only.

a $\quad$ = cross-sectional area, ft^2 (m^2)

A $\quad$ = drainage area, acres (m^2), used in Rational Method

A_m $\quad$ = watershed drainage area, mi^2, used in SCS Method

B $\quad$ = top width of a channel or culvert, ft (m)

c $\quad$ = entrance coefficient (dimensionless) used in orifice equation.

$\quad\quad$ = runoff coefficient (dimensionless) used in Rational Method

$\quad\quad$ = discharge coefficient (dimensionless) used in weir equation

CN = curve number (dimensionless) used in SCS Method

d $\quad$ = distance, ft (m)

d_{50} $\quad$ = median stone size, ft (m), used for riprap

D $\quad$ = diameter of pipe, ft (m)

$\quad\quad$ = normal depth, ft (m)

D_c $\quad$ = critical depth, ft (m)

D_0 $\quad$ = maximum pipe or culvert width, ft (m)

E $\quad$ = specific energy, ft (m)

ε $\quad$ = roughness, ft (m), used in connection with Bernoulli's Equation. (ε is the Greek letter *epsilon*.)

f $\quad$ = friction factor (dimensionless) used in Bernoulli's Equation

F $\quad$ = force, pounds (N)

F_p $\quad$ = pond and swamp adjustment factor (dimensionless) used in SCS Method.

g $\quad$ = acceleration of gravity, 32.2 ft/s^2 (9.81 m/s^2)

γ = specific weight of water, 62.4 lb/ft^3 (9.8 x 10^3 N/m^3) (γ is the Greek letter *gamma*.)

h   = vertical height of water surface above an arbitrary datum, ft (m)

h_f = friction head loss, ft (m)

h_e = entrance head loss, ft (m)

h_L = total head loss, ft (m)

H   = head above crest, ft (m), used in weir equation

HW = headwater depth, ft (m)

i   = rainfall intensity, in/h (m/s), used in Rational Method

I   = flow into a reservoir, cfs (m^3/s)

I_a = initial abstraction or losses, inches, used in SCS Method

k_e = entrance loss coefficient (dimensionless) used with culverts

K   = kinetic energy, ft-pounds (N-m)

L   = length of pipe or culvert, ft (m)

     = length of flow, ft, used in SCS Method overland travel time

     = effective crest length, ft (m), used in weir equation

L'   = measured crest length, ft (m), used with sharp crested weirs.

L_a = apron length, ft (m)

m   = mass, slugs (kg)

n   = roughness factor (dimensionless) used in Manning's Equation

     = roughness coefficient (dimensionless) used in SCS Method overland travel time.

     = number of contractions (dimensionless) used with sharp crested weirs.

N_R = Reynold's Number (dimensionless)

υ = viscosity of water, 1 x 10^{-5} ft^2/s (9.29 x 10^{-7} m^2/s) (υ is the Greek letter *nu*.)

O   = outflow from a reservoir, cfs (m^3/s)

p   = water pressure, pounds/in^2 (N/m^2)

     = wetted perimeter, ft (m)

P   = precipitation, in (m)

     = height of crest, ft (m), used with sharp crested weirs

P_2 = precipitation for 2-year, 24-hour storm, inches, used in SCS Method

q_p = peak runoff, cfs, used in SCS Method. (Note: cfs is abbreviation for ft^3/s)

q_t = unit discharge at time t, csm/in, used in SCS Method (Note: csm is abbreviation for cfs/mi^2)

q_u = unit peak discharge, csm/in, used in SCS Method (Note: csm is abbreviation for cfs/mi^2)

Q   = quantity or rate of flow of water, cfs (m^3/s) (Note: cfs is abbreviation for ft^3/s)

     = runoff, inches, used in SCS Method

Q_p = peak runoff, cfs (m^3/s), used in Rational Method (Note: cfs is abbreviation for ft^3/s)

R   = hydraulic radius, ft (m)

s = gradient, ft/ft (m/m)

   = slope (%)

S = potential maximum retention after runoff begins, inches, used in SCS Method.

ΔS = change in reservoir storage during time Δt, ft^3 (m^3)

t_c = time of concentration, min

T_t = overland travel time, hours, used in SCS Method

TW = tailwater depth, ft (m)

TW'= vertical distance from invert of culvert (at outlet) to hydraulic grade line, ft (m)

Δt = incremental time period, min, for reservoir routing

θ = angle, degrees, made by a v-notch weir

U = potential energy, ft-pounds (N-m)

v = average velocity of a cross section of water, ft/s (m/s)

V = volume, ft^3 (m^3)

W = weight, pounds (N)

   = apron width, ft (m)

z = vertical depth below a free water surface, ft (m)

H

UNIT CONVERSIONS

ENGLISH-METRIC CONVERSIONS

The following table states the equivalence between selected quantities expressed in both the English system and the metric system (S.I). The following example illustrates how to use the table.

Example

To convert 25 cfs to m³/s, first locate the equivalence relation between cfs and m³/s

1 cfs = 0.02832 m³/s.

Then multiply 25 cfs by a fraction consisting of the two equal numbers above with cfs in the denominator:

$$25 \text{ cfs} \times \frac{.02832 \text{ m}^3/\text{s}}{1 \text{ cfs}} = 0.71 \text{ m}^3/\text{s}. \quad \text{(Answer)}$$

Length

1 inch = 0.02540 meter

1 foot = 0.3048 meter

1 mile = 1609 meters

Area

1 square inch = 6.452 x 10^{-4} m²

1 square foot = 0.0929 m²

1 acre = 4046.9 m²

1 acre = 0.40469 hectare

Volume

1 cubic inch = 1.64 x 10^{-5} m³

1 cubic foot = 0.0283 m³

Velocity

1 foot/second = 0.3048 m/s

Force

1 pound = 4.448 N

Pressure

1 pound/square inch = 6895 N/m^2 (pascal)

1 pound/square foot = 4.788 N/m^2 (pascal)

Discharge

1 ft^3/s (cfs) = 0.02832 m^3/s

1 ft^3/s (cfs) = 28.32 l/s

MISCELLANEOUS UNIT CONVERSIONS

Area

1 acre = 43,560 ft^2

1 s.m. = 640 acres

1 hectare = 10,000 m^2

Volume

1 ft^3 = 7.48 gallons

INDEX

computation of, 121, 125
   involving sub-basins, 125
direct measurement, 108
limitations of use, 128
synthetic, 108–109
unit, 108–109, 121–122
hydrologic cycle, 81–82
hydrology, 1–2
hydrostatics, 15

I
impoundment, 249–254, 261–264
   reasons for exceeding the design maximum, 261
infiltration, 82, 97–98, 106–108, 110
initial abstraction, *see* initial losses
initial losses, 107, 110
inlet, 141–143, 145, 147–151
   apron, design of, 202
   control, in culverts, 69–73
   grate capacity, checking, 151
   improved structures, types of, 220–221
   location, factors involved in choosing, 148
   spacing of, 148
intensity-duration-frequency (I-D-F) curve, 96
   usage of, 96
International System of Units (SI), use of, 11
irrigation, 1–3

L
lag, 108

M
manhole, use of, 148
Manning's Equation, 49–53, 89, 198
   parameters used in, 50
metrication, 11
metric system, *see* International System of Units
Moody Chart, use of, 30

N
Natural Resources Conservation Service (NRCS), 5
nomograph, 89
   used to compute flow, 89
normal depth, 46–47, 49, 51, 69, 75–76
   computation of, 46, 52–53
     in pipes, 56
     in streams, 62, 65

O
open channel hydraulics, 45, 48
orifice, 253–254, 256, 258, 260–261
outfall
   configuration, three major types, 164
   design of, 164–173
   maximum velocity of flow through, 168

outlet
   apron, design of, 200
   control, in culverts, 45, 69, 73–75
   multi-stage, 256–261
   structure, 253–261

P
peak discharge, 150–158
piezometer, 19
pipe, 141, 144–146, 148, 150–154, 159–164, 168–171
   capacity, checking, 151
   cross-sections, typical, 144
   materials used in manufacture of, 144
   segment, designing, 145
   size, choosing, 159
   slope, 148, 159
   velocity of flow through, controlling, 145
planimeter, 87
point of analysis, 84–85, 88–89, 97, 107, 110, 121, 124–125, 127
pollution
   of stormwater, 4
   prevention of, 241, 246
pond, used in detention, 241–242, 246
pond and swamp adjustment factor, 118
   not used with hydrographs, 122
ponding, 103, 106, 108, 110
pressure, 17–20
   definition of, 17
   equivalence to potential energy of water, 25
   gauge pressure, 19
   relation to depth, 17
project meeting
   necessary in storm sewer design, 145
   objectives of, 145–146

R
rainfall, 81–83, 92–98, 103, 106–110, 112, 118, 121
   average pattern of, determining, 92–96
   distribution types, 118, 121
   relationship between intensity and duration, 96
rainfall excess, 107–109, 112
Rainfall Frequency Atlas of the United States, 92, 138
Rational Method, 96–103, 106, 109–110, 112, 150–151
   limitations of, 103
   Modified Rational Method, 103, 106
   primary application of, 103
   reasoning behind, 97
   used for determining peak runoff, 96
reservoir, 241, 264
return period, 92, 96
Reynolds Number, 27–28